HISTORY OF TECHNOLOGY

History of Technology

Volume Twenty-five, 2004

Edited by
Ian Inkster

thoemmes

Thoemmes Continuum
The Tower Building, 11 York Road, London SE1 7NX
15 East 26th Street, New York, NY 10010

British Library Cataloguing-in-Publication Data

A catalogue record for this book is available from the British Library

ISBN 0-8264-7187-0
ISSN 0307-5451

Typeset by BookEns Limited, Royston, Herts.
Printed and bound in Great Britain by
The Cromwell Press, Trowbridge, Wiltshire

This volume is dedicated to the memory of Gerry Martin of the Renaissance Trust, engineer, scholar, a man of reliable knowledge.

Contents

Editorial

The bulk of this volume is devoted to a special issue on the steam engine that has been constructed from several papers and responses delivered at an international workshop on that subject held in Windsor in November 2001.[1] A major purpose of this meeting was to utilize detailed considerations of the global history of the steam engine in an effort to isolate the character of the divergence of industrial experiences between East and West that emerged sometime during or before the eighteenth century. Older perspectives stressed that the West had evolved political, institutional or cultural advantages over a considerable period of history, and that this had mushroomed into industrial capitalism. The rest of the world was either incapable of such a transformation, was simply beaten in a race towards it, or was later held back by forces generated by that industrial transition – for instance, by the negative character of European colonialism in Africa and in Asia.

One potentially illuminating way to resolve such arguments might be technological, and might emerge from a reconsideration of the history of the steam engine. Although its importance as a prime mover may well have been relatively slight for some time, few historians have doubted the power of steam to promote other crucial elements of the industrial revolution, including the essential move to mineral energy feedbacks to metallurgical industries and the turn towards innovations in machinery involving finer instruments of measurement and materials quality, gearing, power transmission and novel machine tools.[2] Thus, general global claims could be specified as a more exact query: could Chinese civilization have come up with a steam engine comparable with that improved and perfected in Europe during the eighteenth and nineteenth centuries? If not, may we fasten on cultural, scientific, institutional, political or economic reasons for contrasts between East and West? This, of course, could never be the only form of the enquiry, for just as pertinent to the history of our globe might be: was there something in the system of western industrialism that prohibited the Chinese from successfully adopting the western steam engine into manufacturing production prior to, say, 1911? Nevertheless, on offer in this special issue of the journal is a form of historical enquiry that serves to focus the mind, and the papers and discussions elaborate upon a variety of answers or types of answer, ranging from the isolation of a missing 'factor' in China, to the notion that China and the West were on differing economic and institutional trajectories.

The three papers in this volume of the journal that precede the Special Issue, themselves represent a good range of the history of technology. The paper by Nick Hayes moves towards a social history of technology through examination of the institutional and local contexts of low-rise systems of housing in Britain after 1945. A conclusion is that the development of such new systems was hampered by contending interests rather than by technical lacunae as such. The article by Watts and Langdon is a return and corrective to a paper first published in this journal in 1992. The narrative represents a reconstruction based on minute textual examination and a logic that draws upon precise knowledge of the materials, skills and wages, and practices of the early fourteenth century. In this paper the limits of textual evidence are the point at which archaeology begins. The third paper, by Jan van den Ende of Rotterdam, argues that recent social-systems approaches to technology have in fact obscured the impacts of technology on society. In a paper on twentieth-century computing the author suggests the relevance of this to the history of technology.

Notes and References

1. 'The Evolution and Diffusion of Steam Power and Stream Engines in Europe compared with China from 1589 to 1914', Third Windsor Conference on the Global History of Material Progress, Cumberland Lodge, Windsor Great Park, 15–17 April 2002.

2. None of which is to say that other comparative technological devices could not have been chosen. For an early example of just this see Joseph Needham, *The Development of Iron and Steel Technology in China*. The Second Biennial Dickinson Memorial Lecture, Newcomen Society, Science Museum, London, 1958.

The Contributors

Prof H. Floris Cohen
Raiffeisenlaan 10
3571 TD Utrecht
The Netherlands

Dr Kent G. Deng
LSE
Houghton Street
Aldwych
London WC2A 2AE

Prof Mark Elvin
Division of Pacific and Asian
    History
Research School of Pacific and
    Asian Studies
ANU, 9
Canberra
Australia ACT 0200

Dr Nick Hayes
Humanities Faculty
Nottingham Trent University
Clifton Lane
Nottingham NG11 8NS

Dr Richard Hills
Stamford Cottage
47 Old Road
Mottram
Hyde
Cheshire SK14 6LW

Dr Graham Hollister-Short
119 Park Road
Chiswick
London W4 3EX

Dr Rob Iliffe
Centre for the History of Science,
    Technology and Media
Imperial College
South Kensington
London SW7 2AZ

Prof Ian Inkster
Humanities Faculty
Nottingham Trent University
Clifton Lane
Nottingham NG11 8NS

John Langdon
Department of History and Classics
University of Alberta
2–28 Henry Marshall Tory Building
Edmonton, Alberta
Canada T6G 2H4

Dr Liu Chun-Yu
Department of English
Wenzao Ursuline College of
    Languages
900 Mintsu 1st Road
Kaohsiung 807
Taiwan

Dr Alessandro Nuvolari
Eindhoven Centre for Innovation
Studies ECIS
Faculty of Technology Management
TEMA 1.14
Eindhoven University of Technology
PO Box 513 5600 MB Eindhoven
The Netherlands

Patrick O'Brien
LSE
Houghton Street
London
WC2A 2AE

Prof Nathan Sivin
History and Sociology of Science
University of Pennsylvania
Philadelphia
PA 19104-6304
USA

Prof Jan van den Ende
Rotterdam School of Management,
F2-67
EUR
PO Box 1738
3000 DR Rotterdam
The Netherlands

Martin Watts
1 Trinity Cottages
Cullompton
Devon EX15 1PE

Prof R. Bin Wong
Department of History 200 HOB
University of California, Irvine
Irvine CA 92697-3275
USA

Dr David Wright
118 Bathurst Road
Winnersh
Wokingham
Berks RG41 5JF

Notes for Contributors

Contributions are welcome and should be sent to the editor. They are considered on the understanding that they are previously unpublished in English and are not on offer to another journal. Papers in French and German will be considered for publication, but an English summary will be required. The editor will also consider publishing English translations of papers already published in languages other than English. Include an abstract of 150–200 words.

Authors who have passages originally in Cyrillic or oriental scripts should indicate the system of transliteration they have used. Be clear and consistent.

All papers should be rigorously documented, with references to primary and secondary sources typed separately from the text, double-line spaced and *numbered consecutively*. Cite as follows for:

BOOKS

> 1. David Gooding, *Experiment and the Making of Meaning: Human Agency in Scientific Observation and Experiment* (Dordrecht, 1990), 54–5.

Only name the publisher for good reason.

Reference to a previous note:

> 3. Gooding, *op. cit.* (1), 43.

Titles of standard works may be cited by abbreviation: *DNB*, *DBB*, etc.

THESES

Cite University Microfilm order number or at least Dissertation Abstract number.

ARTICLES

> 13. Andrew Nahum, 'The Rotary Aero Engine', *Hist. Tech.*, 1986, 11: 125–66, esp. 139.

Please note the following guidelines for the submission and presentation of all contributions:

1. Type your manuscript on good quality paper, on one side only and double-line spaced *throughout*. The text, including all endnotes, references and indented block quotes, should be in one typesize (if possible 12 pt).

2. In the first instance submit two copies only. Once the text has been agreed, then you need to submit three copies of the final version, one for the editor and two for the publishers. You should, of course, retain a copy for yourself.

3. Number the pages consecutively throughout (including endnotes and any figures/tables).

4. Spelling should conform to the latest edition of the *Concise Oxford English Dictionary*.

5. Quoted material of more than three lines should be indented, without quotation marks, and double-line spaced.

6. Use single quotes for shorter, non-indented, quotations. For quotes within quotes use double quotation marks.

7. The source of all extracts, illustrations, etc., should be cited and/or acknowledged.

8. Italic type should be indicated by underlining. Italics (i.e. under-lining) should be used for foreign words and titles of books and journals. Articles in journals are not italicized but placed within single quotation marks.

9. **Figures.** Line drawings should be drawn boldly in black ink on stout white paper, feint-ruled paper or tracing paper. Photographs should be glossy prints of good contrast and well matched for tonal range. Each illustration must be numbered and have a caption. Xerox copies may be sent when the article is first submitted for consideration. Please do not send originals of photographs or transparencies but if possible have a good-quality copy made. While every care will be taken, the publishers cannot be held responsible for any loss or damage. Photographs or other illustrative material should be kept separate from the text. They should be keyed to your typescript with a note in the margin to indicate where they should appear.
   Provide a separate list of captions for the figures.

10. Notes should come at the end of the text as endnotes, double-line spaced.

11. It is the responsibility of the author to obtain copyright clearance for the use of previously published material and for photographs.

An Early Tower Windmill?
The Turweston 'Post mill'
Reconsidered

MARTIN WATTS AND
JOHN LANGDON

In the 1992 volume of this journal, a case study was provided of the
construction and subsequent operation of a windmill built on the
Westminster Abbey manor at Turweston in the Midlands of England in
1303.[1] It was felt at the time that this was a post mill with a low stone
support wall around it. However, discussion between the two authors,[2] the
discovery of a mistake in the translation of the construction account,
comparison of the Turweston mill with the nearly contemporary tower
windmill at Dover (built 1294–5), and the recent appearance of useful
comparative material about medieval Continental tower mills has all
prompted a reinterpretation of the evidence concerning the Turweston
windmill. In short, it is now felt by us that the Turweston windmill was
possibly an early tower windmill or at the very least a composite, mixing
post- and tower-mill forms. If so, it adds a significant new example, not
only to the number of possible early tower windmills but also as a further
indication of medieval technical creativity.

The key piece of evidence for elucidating the form of the windmill is still
the 'Cost of Mill' section found in the 1302–3 manorial account roll for
Turweston,[3] detailing the construction of the windmill, a translation of
which was published in the appendix of the above 1992 article.[4] The first
part of that section is taken up with a detailed description of the
construction of a 'wall' (*murus*) for the 'new windmill'. It was of rubble
masonry construction, consisting of a combination of lime mortar and
loose stone, the latter dug up from the surrounding countryside (as
indicated in the first two entries in the construction account). This 'wall'
was eventually topped with a *tabulamentum*, or a course of finished stone. It
was originally thought that this was a low wall to protect the base of a post
mill, a view based primarily upon the fact that apparently only five
cartloads of stone were used in the construction of the 'wall' (from the first
entry, as in the published 1992 translation). When the document was
revisited as a result of the above-mentioned discussion by the authors,
however, it was found that the number of cartloads was 500, a result of

misreading a small abbreviation mark in the document.[5] This indicates
that the 'wall' was much more substantial than originally thought and
might certainly have been high enough to constitute a tower.[6] A number of
calculations performed by the authors, converting the 500 cart-loads of
stone into their equivalent volume of masonry rubble[7] and assuming
tower-mill dimensions at the base similar to those from surviving remains
of medieval or early modern tower windmills,[8] have suggested that the
Turweston 'wall' could have been up to 30 feet high and was probably at
least 20 feet, certainly tall enough to accommodate the seemingly 18–22$\frac{1}{2}$
foot sail yards.[9]

   Such figures help to reconcile what were thought in the 1992 article to
be rather puzzling features of the windmill's construction. For example,
the two brass wheels, one 'brass', and the eight strips of brass bought for
the wooden wheels for the large sum of 29s. 7$\frac{1}{4}$d. now make more sense as
probably referring to the rollers and brass supporting strips upon which a
tower-mill cap revolved.[10] Similar language also occurs in the construction
account for the almost contemporary tower windmill built at Dover Castle
in 1294–5, where the wheels upon which the Dover tower-mill cap rotated
were called *ruelli* (literally 'disks') and were seemingly affixed with iron
pins (*kuvillae*; from *cavilla*) to the top inside edge (presumably) of the
tower.[11] The terminology in both the Dover and Turweston cases also
parallels that later used in the construction of Flemish tower mills, where
the French *rondielles* or *molettes* signified the rollers and *courbes* the curved
pieces of wood between which the rollers were set.[12] Finally, the various
later references to *le carole* of the Turweston mill, with its implications of a
circular shape or movement,[13] become more explicable if it is seen as being
the rotating cap of a tower mill rather than the support structure for a post
mill, as originally thought.[14] Other details are also possibly indicative of a
cap, such as the reference in the 1325–6 Turweston account to sawing
timber for *le Schroud* of the *carole*.[15] This might be roofing of the cap itself or
some sort of boarded cladding covering the gap between the cap and the
tower, a feature known later in tower mills as a 'petticoat' or 'skirt'.
Similarly, the curious reference to 8d. spent on lead for guttering in the
original building account makes more sense in relation to tower mills,
where use of lead for guttering and flashing was common.[16] Altogether, if
these lines of argument are accepted, the mill might have looked
something like the tower mill shown in Figure 1a.

   Nevertheless, there are a number of troubling points about the
construction of the mill and its subsequent operation which are not
wholly consistent with it being a tower mill. For example, at no time
throughout the extensive documentation for the mill was it called a
'windmill of stone' (*molendinum ventriticum de petra*) or 'stone windmill'
(*molendinum ventriticum petrinum*), as medieval tower mills were more
normally designated.[17] This may be because the terminology had not
yet circulated sufficiently to have become commonplace by 1303, but the
fact that the Dover mill was called a 'windmill of stone' nearly ten years
before would seem to make this a questionable supposition. Another

a

b

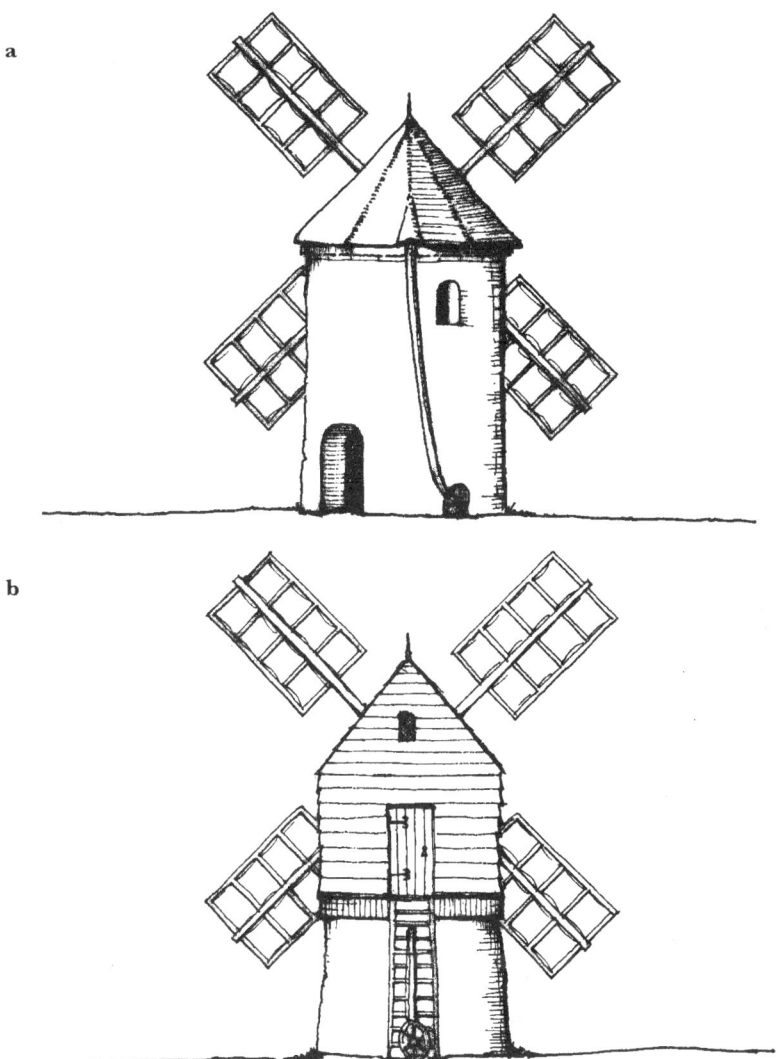

Figure 1

puzzling feature is the small amount of cash actually spent on the Turweston 'wall' ($£5$ 4s. $11\frac{1}{4}$d. or 26.5 per cent of the $£19$ 16s. $3\frac{1}{2}$d. total spent on the mill as a whole). This was far less than that spent on the Dover tower ($£18$ 7s. 2d. or 50.5 per cent of the $£36$ 6s. 11d. total). Similarly, the Turweston 'wall' only took six weeks to build compared to the *minimum* of ten weeks needed for Nicholas of Aynho, the master mason, and his team of five or six other masons to build the tower for the windmill at Dover.[18] Nor does the 35s. cash payment (plus one quarter of wheat) for building the Turweston 'wall' by an unspecified number of workers seem very high, and indeed if the workers were paid the normal craftsman's wage (at the time) of 4d. per day then it would suggest only three of them were engaged over the six-week period.[19] Comparison with other cases of rubble masonry wall construction at the time suggest that wages of 35s. would not be sufficient to build a circular wall of even 10 feet high at the diameter required for a tower mill.[20]

There are other pieces of evidence suggesting that the Turweston stone work was built to a relatively low height. Unlike the Dover mill, for instance, where at least two wooden floors were built inside the tower (making three storeys in all), there was no sign of this in the Turweston case, the only reference being to 'frames' made for the 'steps and stairway' to the mill. This stairway may have been installed inside the tower, or possibly even outside it, leading directly up to the wooden superstructure, if that superstructure were not too far off the ground. It is notable, too, that the Turweston mill account only ever alludes to one door,[21] contrary to the Dover case where 'doors' are mentioned. Multiple doors, in fact, were normally found on tower mills, since at least two doors or entry points were needed so that access to the mill was always possible no matter from which direction the wind was blowing.[22] Nor was there any allusion to corbelling for doors or windows in the Turweston 'wall' as there was for the Dover tower. Finally, there was a fairly hefty expenditure of wood for the Turweston mill, which seems excessive for the cap of a tower mill, and a more substantial superstructure on the 'wall' might have been likely.[23]

In short, the Turweston material supplies none of the clear information regarding the internal layout of the tower that the Dover tower mill in particular supplies, and in fact we might be looking at a relatively short stone foundation, up to say 10 feet high, with no openings for doors or windows and perhaps with relatively thick walls (perhaps up to 4 feet or more, which would help to explain the many cartloads of stone used), in order to support a substantial post-mill-like body rotating on rollers, a possible version of which is shown in Figure 1b. In this case, the *carole* mentioned above might be the turntable upon which the post-mill body rested. This would help to explain the reluctance to call the mill as a whole a 'stone windmill', the lack of references to the central 'post' or 'standard' upon which normal post-mills swivelled, and perhaps even the initial difficulties the miller seems to have had in turning the mill, since an arrangement of this type would put a lot of weight on the rollers.[24] Unfortunately no obvious signs of any sort of windmill remain on the landscape today to help us judge the matter,[25] and a more intensive geophysical survey or even archaeological excavation will probably be necessary to reveal significantly new information.

The frustratingly ambiguous nature of the terminology used for constructing and maintaining the windmill as recorded in the Turweston manorial accounts might well mean that we shall never be completely sure of the exact nature of the mill, even with the help of archaeology. One thing, we think, is certain – that the Turweston windmill was not a post mill in the commonly accepted sense. It might have been an early tower mill, or possibly a composite mill mixing post- and tower-mill elements. Both possibilities speak to a notable medieval inclination for technical experimentation. The business the windmill did in its first two months of operation suggests that it was quite a novelty.[26] It was not a sound investment, however, and was probably too expensive an enterprise for the clientele it served, such that it was derelict by 1338.[27] As in our own time, introducing new technology was a risky business for medieval people. That they were still willing to do it is an impressive comment on the age.

Notes and References

1. John Langdon, 'The Birth and Demise of a Medieval Windmill', *History of Technology*, 1992, 14: 54–76.

2. In early 2002 Watts e-mailed Langdon with the suggestion that the Turweston mill might in fact be a tower windmill. A lively 'virtual' discussion ensued, which eventually persuaded Langdon to revisit the documentation and to reconsider his position.

3. Westminster Abbey Muniments (henceforward WAM) 7764.

4. Langdon, *op. cit.* (1), 69–71.

5. Langdon had originally thought that the abbreviation over the 'v' (the Roman numeral for 'five') was a 'q$\mathfrak{z}$', that is, indicating the Latin *quinque*, which form often occurs in the documents. But, with the use of a magnifying glass, the abbreviation is clearly a 'c', that is, indicating the 'five' should be multiplied by a hundred. As a result of discovering this mistake, Langdon has re-examined the documentation and calculations for the 1992 article. Two other smaller mistakes were found in the construction account. The entry (p. 69), 'In sand dug up for the same [windmill], 7s. 3$\frac{1}{4}$ d.' should read 'In sand dug up for the same [windmill], 7s. $\frac{3}{4}$ d.', while on p. 70, the entry 'In expenses made concerning the carriage of timber ... 1 quarter and 1 bushel of wheat [and] 6$\frac{1}{2}$ bushels of oats' should read 'In expenses made concerning the carriage of timber ... 1 quarter and 1 bushel of wheat [and] 1 quarter [and] 6$\frac{1}{2}$ bushels of oats'. There were also some computational errors in Table 1 of the 1992 article (p. 62), where the 'Grain received as multure' figures for 1307–8, 1308–9 and 1309–10 should be 64.6 (instead of 92.5), 59 (instead of 83) and 63 (instead of 91) respectively. All other figures in the 1992 article appear to be correct. Langdon apologizes to the readers of the journal for these errors.

6. The word *turris* (Latin for tower) was never used in relation to the Turweston mill, but neither was it for the undoubted tower mill at Dover Castle, Kent, built in 1294–5, where again the word *murus* was used throughout in describing the building of the windmill tower: Public Record Office (hereafter PRO), Kew, London E101/462/14.

7. Here it was assumed that each cartload could carry a minimum of one-half imperial ton of stone, the maximum probably being around three-quarters of a ton: e.g., John Langdon, 'Horse Hauling: A Revolution in Vehicle Transport in Twelfth- and Thirteenth-Century England?', in *Landlords, Peasants and Politics in Medieval England*, ed. T.H. Aston (Cambridge, 1987), 63–4. The composition and density of limestone rubble mortar was taken from modern engineering data given in *Perry's Chemical Engineer's Handbook*, 7th edn, Robert H. Perry and Don W. Green (eds.) (New York, 1997), 2–119.

8. As at Burton Dassett (Warwickshire), Tidenham (Gloucestershire) or Fowey (Cornwall): Martin Watts, *The Archaeology of Mills and Milling* (Stroud, 2002), 111–13. All these mills had an external diameter of around 20 feet.

9. As calculated from the lengths of sail cloth bought each year: Langdon, *op. cit.* (1), 74, n. 37.

10. Langdon, *op. cit.* (1), 70.

11. PRO E101/462/14, m. 3.

12. Yves Coutant, *Windmill technology in Flanders in the 14th and 15th centuries: Part 1: The external structures of early post and tower mills*, translated [from the French] by Michael Harverson and Owen Ward (International Molinological Society, vol. 16, Watford, 2001), 76–7, 79–82. *Curbae* were mentioned in relation to the rollers in the Dover account, as when a carpenter was

hired for working *circa curbas & ruellos* (PRO E101/462/14, m. 3), but no *curbae* were referred to in the Turweston account.

13. E.g., Frederic Godefroy, *Dictionnaire de l'Ancienne Langue Française*, 10 vols (1937–8 edn, Paris), i. 786.

14. Langdon, *op. cit.* (1), 59, 67.

15. WAM 7781.

16. Langdon, *op. cit.* (1), 70; Coutant, *op. cit.* (12), 83–4.

17. E.g., Coutant, *op. cit.* (12), 72, 76–7.

18. Nicholas and five masons worked on the tower for two weeks in late November and early December and then, after a break because of winter, he continued work on the tower with six masons over eight weeks from late March to late May (with a break of one week at Easter). This does not include the many weeks Nicholas spent cutting stone in preparation for this work plus a week (after a break for Pentecost) that he spent with two masons in late May–early June in finishing off the tower. In all of this the masons were assisted by a bevy of general labourers, both men and women, who mixed mortar, fetched water (the women's work) and carried stone 'to the hands' of the masons: PRO E101/462/14.

19. Langdon, *op. cit.* (1), 69. Three men working at 4d. per day for a six-day working week would, over six weeks, be paid 36s., very close to the 35s. actually expended. Perhaps the quarter of wheat was given in place of the missing shilling.

20. For example, at Hollingbourne, Kent, in 1312–13, an enclosing wall of rubble masonry (that is, again loose stone and lime mortar) was erected around a watermill to a height of nine feet at a labour cost of 9s. 6d. per perch: Canterbury Cathedral Archives DCc Hollingbourne 25. A perch was probably around $16\frac{1}{2}$ feet in length at this time, so that the area of this perch of nine-foot-high wall would be 148.5 square feet. If the Turweston tower 'wall' was built for the same amount per square foot, then the 35s. for the Turweston mill would buy the construction of 547.1 square feet. To estimate what this would mean in terms of height, we assume a mean diameter of, say, 18 feet between the external and internal diameters of the Turweston tower (about normal for early tower mills), the circumference would be 56.5 feet. Dividing this circumference into the 547.1 square feet above (and assuming that the tower was parallel-sided) would yield a height of only 9.7 feet.

21. Langdon, *op. cit.* (1), 66, 70.

22. If only one door was available, the entrance might on occasion be blocked by the windmill sails: Watts, *op. cit.* (21), 111–12.

23. Nearly a ton and a half of boards (103 'quarters' at probably 28 lbs per 'quarter') were bought in addition to timber obtained from the market town of Bishop's Itchington or from the Abbey's manor at Knowle, both in Warwickshire: Langdon, *op. cit.* (1), 70. Comparison with other windmill construction accounts for post mills suggests that the wooden superstructure for the Turweston mill was as substantial as those for post mills. For instance, £8 7½d. was spent on the wooden superstructure of the Turweston mill compared to £9 15s. 2d. for the superstructure of the post mill built at Walton, Somerset, in 1342, or the £5 15s. 9½d. spent on the superstructure for the post mill built at Burstwick, Yorkshire, in 1296–7: Langdon, *op. cit.* (1), 70–1; Ian Keil, 'Building a Post Windmill in 1342', *Transactions of the Newcomen Society*, 1961–2, 34: 151–4; PRO SC6 1079/15, m. 7v. (These amounts include all timber and carpentry expenses along with smithy costs for nails, hooks, hinges, iron collars, plates, etc. They do not include millstones, canvas or, for the most part, the internal workings of the mill.)

24. See Langdon, *op. cit.* (1), 60, for this last. The problem of moving the mill was seemingly solved by attaching a wheel to the end of the tail-tree and laying down a stone path upon which this wheel would run as the mill superstructure was turned, so that the sails would face the wind.

25. The authors examined the likely site for the windmill in the summer of 2002 (see Langdon, *op. cit.* (1), 57, for a map). There were a few low hummocks in the fields (of pasture) inspected, which might conceal the base of medieval or post-medieval windmills, but without extensive excavation this must remain conjectural.

26. Langdon, *op. cit.* (1), 61–2.

27. Langdon, *op. cit.* (1), 62–8.

Prefabricating Stories: Innovation in Systems Technology after the Second World War

NICK HAYES

Commenting in 1944 on the future design of housing for post-war reconstruction, the British government's Central Housing Advisory Committee, an eclectic body of expert talents, noted that:

> The process of housing construction is developing in the direction of the greater preassembly of parts of the house at the factory. It is not yet possible to state with any confidence how far such methods can be carried with satisfactory results. While, therefore, the case for entire prefabrication is by no means established, it is possible that in the future complete houses may come to be built in this way.[1]

How should we read such a statement? Stories about buildings, and thus their social meaning, vary considerably through time.[2] Without knowledge of the 'high-rise' systems débâcles of later decades, we might wonder at CHAC's cautious optimism. Yet we might ponder, too, that during the inter-war years site methods had shifted only 'very slowly from making to assembling'.[3] Learning that earlier experiments with prefabricated houses had failed technically, were more expensive than traditional methods, and were remembered unfavourably by enthusiasts and consumers alike, acting as contemporaries we might then dismiss such forecasts as fanciful and self-deluded.[4] Indeed, on being harangued by *Picture Post* (the exemplar advocate of reconstruction) one month later that 'we must work out some totally new way of building' – demand the 'prefabricated house' – we might rightly conclude that we had entered a looking-glass world.[5]

Technology is a social product: understanding the social dimension, it is argued, is crucial to our understanding of its success or failure.[6] Housing, especially, has a close and intimate social meaning. Of course, most of us have specific, personalized understanding(s) of everyday things: our homes included. What we understand, individually or as groups, we understand through different filters, for different reasons, and with different priorities. Thus, while technologists spent the war years enthusiastically investigating building systems utilizing steel and concrete that used less skilled site labour to construct the external shells of houses, for tenants physical structure remained only immediately significant if it 'was defective and let the rain through'. Contextually, anyway, most householders were 'incapable of imagining the sort of world where they would be allowed to choose the home they wanted'. External appearance for housing authorities, however, was important; new housing stock ideally had to marry with existing.[7] And planners, architects and politicians – experts generally – it is argued, also had different group visions: lest we too readily conflate the widespread popular interest in housing at a time of acute shortages with shared professional preoccupations.[8]

Yet the limits and range of shared social meanings are important to understanding technological development.[9] Meaning through interaction brings agency, direction, cohesion. It sets markets, creates networks and stamps trajectories. But how is meaning attributed? First, there is a story of internal coherence, the basis on which we judge buildings or other products in themselves, how we comprehend their purpose – aesthetic, functional, economical, etc. Second is the story of their external explanation, of causes and context – 'broader meaning'.[10] Both components have strong human agency: of commitment, rationality, necessity, expense, scarcity, etc. This article explores the interactive nature of 'shared meaning' across this divide between social groups, and the impact this had on the development of low-rise systems housing after 1945. It will suggest that this 'new' technology was widely and continually misunderstood and interpreted by different groups in very different ways, limiting agreement, product stabilization and closure.[11] This proved inimical to subsequent development.

ACTORS AND CONVERSATIONS

Writing in the mid-1960s, Marian Bowley noted that 'innovations in structures and methods of design' had 'developed very tardily in this country': evidenced by 'woefully little general interest in research', in 'economic rationality in design', or in offering customers choice.[12] Particularly important here had been the separation of the design and construction functions, so neither architect nor builder had the incentive or authority to conjure innovation. Even greater explanatory weight has subsequently been assigned to this 'social' dimension of organizational interaction.[13] If 'successful innovation requires the coupling of the technical and economic, rather than being solely a matter of "technology push" or "market pull"', then inter-functional co-operation and communication offers the very force for change.[14] This is perhaps even truer of construction, as essentially an assembler of other manufacturers' products: 'really just a network and a set of lists, like a telephone directory', perhaps with 'too many characters and no plot'. In this anarchic system, 'common meaning' and 'common understanding' were therefore at a premium;[15] the 'quality' of conversation (that is, information and knowledge) between actors having a significant impact on the force and direction of innovation.[16]

Yet supplier-dominated concerns like construction, it is argued, 'appropriate less on the basis of technological advantage, than of professional skills, aesthetic design,' etc.:[17] where design is seen as more an art than a science. If this was true of traditional 'bricks and mortar' construction, a sector exemplifying small, low-visibility and cumulative change, was it also true of non-traditional, systems forms of construction?[18] Contemporaries argued not. As one prominent building economist noted at the end of the war, there was 'some reason for believing that at last ... a genuine revolution in technique' was about 'to take place'. Scientific discourse, too, was 'swinging against ... traditional methods' towards non-traditional methodologies.[19] Such beliefs were certainly advanced by architectural theorists on other than aesthetic grounds (for example, in terms promoting 'modernizing' industrial efficiency). These discourses co-existed with increased technological investigation into new building methodologies. Each gave the other validity and vitality, even if the claims made were questionable.

Beyond this, did non-traditional networks conform to the same 'conversational' patterns between actors as in the industry generally? In public housing contracts the customer was 'not an individual' but a 'complex system of differing interests'. 'These client systems ... consisted of both congruent and competing sets of understandings, values and objectives'.[20] Architects provided a powerful professional social filter: as intellectual initiators, and thus as interpreters of the client's fluid demands. Historically they also led the building industry: from the architects' viewpoint, the construction process 'was hierarchical both socially and in terms of working organisation'.[21]

Yet local authorities frequently did not employ architects directly. Moreover, a clear majority of the low-rise housing systems introduced after 1945 was sponsored by building contractors or component suppliers.[22] Did this mean that architects – as professional gatekeepers – were less influential in determining outcomes? A preliminary review would suggest not. Non-traditional housing was almost exclusively a public-sector activity, where tenants' views (directly or indirectly) were not actively sought. 'Consultation' consisted of numerous educative exhibitions and a plethora of pamphlets and books to transmit an imposed experience. The 'community' anyway, apparently, was less interested in 'planning', etc., than more immediately in 'houses at any price'.[23] Neither were local authorities, as purchasing agents, actively consulted. Instead, they were invited to inspect and consume, advised by ministry architects, but otherwise largely excluded from the early decision-making process.[24] Thus, the 'design of social housing' was controlled 'by the makers of social policy'.[25] It was they who instituted a 'way of thinking' about future problems and what was possible, possessing and applying a 'set of knowledge' and 'expertise' that gave authority and direction.[26] Attention, therefore, has centred on architects as the instigators of systems

methodology, although others – such as politicians, government departments and national housebuilders – are also cited as co-correspondents.[27]

From this, the historiography focuses on the role of ideas and myth; the
binding together of mutual interests utilizing a 'misleading' hegemonic
discourse.[28] Here rationality is subverted and prefabrication is chosen over
traditional forms of construction despite being more expensive, flawed and
less popular with tenants. Actor innovators, anyway, do not always treat
consumer preferences as 'unalterable structural constraints' on their own
behaviour. Technological trajectories can accord more with speculative
'early promises' than hard information.[29] In this respect, of primary
importance in determining trajectories are the *a priori* beliefs held by actors
– their problem-solving rules, specific knowledge, etc., and the social/
environmental hierarchy that determines which beliefs are privileged at a
particular time.[30] In 1940s Britain politicians and housing experts, it is
argued, came to believe that a 'technical breakthrough' in housing was
necessary and had actually occurred, allowing them political and
economic freedoms in terms of production that otherwise would have
been closed because of acute factor shortages.[31] Rationality and
irrationality, it can thus be argued, existed in many forms when socially
or politically contextualized. The pricing mechanism, by itself, although
important, offered only one check for decision-makers. Others at the time
included gross national labour-saving potential, or the need to maximize
housing output (from either a governmental or commercial standpoint).
Thus, what was rational from one viewpoint, was irrational from
another.[32]

If, therefore, common understanding between producer and consumer
was only partially present (in that needs, expectation and promises
overlapped), between producers (that is designers and builders) and
policy-makers (politicians and state servants) a greater congruence existed.
But was this sufficient? In the first decade after the war, the state licensed
all new building work. Thus it had a controlling hand. In addition little
'hard evidence' existed against which to test the claims being made for
non-traditional techniques, although early scientific evaluation did
validate initial belief systems. Yet there was an element, too, of wishful
thinking in such assessments.[33] Indeed any reliance on myth suggests that
ambiguity existed amongst and between actor clusters. Pinch and Bijker
argue that the technical stabilization of a product – that is the closure of
controversy – occurs not when all problems are solved, but when 'relevant
social groups *see* the problem as being solved'.[34] The remainder of this
article will explore whether this occurred.

PREFABRICATION: MORE IN THE MIND?

Interviewed in 1944 on his return from a fact-finding tour of the USA, Sir
George Burt stated that 'if the Enquiry dealt with one point, it established
that so far as prefabrication is concerned it exists in America more in the
imagination than in practice'. Asked then whether the setting up of

factories to produce prefabricated shells and structures in Britain would not assume continuous demand for years, he advised that:

> I think that this is one of the principal stumbling blocks so far as this country is concerned. I have seen no prefabricated house which will be so popular as to create a home market which will justify the somewhat elaborate equipment necessary. In the immediate post-war period it may be necessary to use a prefabricated outside structure to save site man-hours ... [but] I can see no permanent future for such structures.[35]

Burt was not only a prominent national contractor; he also chaired the government committee then investigating alternative methods of house construction. But his comments are contextually significant for other reasons. 'Modern' England, as Priestley recognized in 1934, was already heavily overlaid by American influences.[36] And, in terms of prefabrication, historians have stressed the impact that the idea of America had in shaping the British psyche.[37] It is easy to see why, when *Picture Post*, for example, argued that 'Prefabrication had Cut Down America's Housing Problem', prefabricated houses were erected 'in a Day', and 'Your whole house may be erected in a factory and trucked to site, as has been done in America ... Millions of us will have houses like this'.[38] Nor was the architectural press exempt from this hyperbole. *Architectural Review*, an early advocate of modern design, argued that success in America 'hands us a sharp jolt on the chin'. Focusing on 'prefabrication', and 'the complete factory assembly', it went on to claim that 'The phenomenal U.S. housing output is to be interpreted as an omen of the future ... It shows that it is possible to produce houses quickly, with a high standard of performance and with excellent equipment.'[39]

Yet there was nothing extraordinary about these discourses. Even stripped of its transatlantic focus, to the wider British audience such 'adverts' spoke of what might be obtainable through British technical ingenuity if only ... (just as British technical know-how had triumphed during the war).[40] Indeed, perhaps the semi-illusory quality made the offer more attractive and magical. However, a fascination with what was 'possible' – a mutually accepted interpretation of what technology might provide – needs to be grounded against the public's 'deep anxiety' about housing shortages, which formed the political backdrop.[41] Commenting on public reaction to an exhibition on post-war home designs, Mass Observation, the influential pioneering social-research organization, recorded that:

> People just wondered when all these things would be available, whether they would really be available at all. As one young middle-class woman, married and homeless, said, 'They could just give me any of it, and I should think it wonderful. Honestly, I liked it all. I'm so desperate for a home I'd like anything. I can't criticise or judge it at all – four walls and a roof is the height of my ambition.'[42]

The 'very nice if we ever get it', 'very good, but the point is, how long will it take before they carry it out?' mentality was evident elsewhere.[43] Architectural commentators, judged by the plethora of validatory articles on prefabrication then appearing in the technical press, were significantly ahead of the profession or political opinion.[44] The Royal Institute of British Architects had apparently been 'ostrich-like' in its pronouncements; even in the last months of the war it was still writing reports 'from a position deep in the rear'. The press noted, too, that 'prefabrication had been too much on the defensive'. Mistakenly, prefabricated houses were still not 'put forward as a better method of building than brick and mortar, but simply as a method of building houses without employing bricklayers and plasterers.'[45] 'We hardly yet know how to use it', opined the *Architects' Journal* at the end of 1945: 'Laboriously we imitate the old brick house in form and content. Our prefabricated efforts are mere copyism, their explosive potential still unrevealed.'[46]

A consensus was lacking, therefore, over what should be valued or produced. Early reports from the secretariat of the Burt Committee noted that:

> no proposed method satisfactorily solves the problem of producing houses cheaply. The methods suggested are complicated. I think this is due to the fact that the simpler methods of improvement on the basis of existing practice have been exhausted, and that attempts are being directed on entirely new methods, not yet properly directed.[47]

Others were harsher. The Ministry of Works Deputy Chief Scientific Advisor commented on existing inter-war systems that:

> Many alternatives were tried, and in one respect or another were found wanting. Some of the houses then produced turned out to be thoroughly bad; others proved more costly than the traditional houses, and as a result the alternative systems in use [by 1939] could be numbered on one hand and none was working on any large scale. If we consider the state of scientific knowledge in 1920, it is not a matter of surprise that the alternatives used were unsuccessful. There was no basic scientific data concerning the properties of and behaviour of building materials except for a limited and very narrow fund of knowledge of their strength.[48]

It is not surprising, therefore, that the Burt Committee's initial temper was very cautious, evaluating and reporting on inter-war practice rather than looking to current and future research.[49] Yet, as the *Architects' Journal* commented, what purpose did this serve? A 'far more valuable contribution at this point would be a report, less unadventurous than this first, on entirely new experiments – especially those concerned with prefabrication'.[50]

If architects and builders remained uncertain of what or how to build, the community seemed equally bemused by non-traditional design. One engineer, for example, reporting on 'overheard conversations ... at recent and numerous Housing Exhibitions', recalled that 'misunderstanding and prejudice are about equally mixed in the public mind'. This was the product of 'cellophane wrapped publicity campaigns' and the 'deeply rooted ... acceptance of brick-wall and pitched roof traditional construction ... as the ultimate in building technique'.[51] Certainly, public responses tended to reinforce this view. For example, numbers of people were 'definitely put off' by certain non-traditional types (for, example the Orlit and Unibuild) 'simply because the roof was flat'. Yet the Orlit was also the most popular of house types then being displayed by the Ministry of Health. At least it looked 'permanent', 'attractive', 'spacious'. Steel and asbestos-cement clad houses, by contrast (the British Steel and Braithwaite), were viewed negatively, as being 'cheerless', 'cold' and 'barracky' – 'imagine rows and rows of them'. Indeed, on all non-traditional types, 'by far and away the most unpopular comment was on the external appearance of the houses'. Within a 'comparatively low level of positive interest' generally, a 'quite like them if I could get nothing else, but they've no individuality' view dominated in the public mind.[52]

Figure 1 The Orlit House: the structural frame is of pre-cast reinforced concrete beams and columns connected on site to produce a continuous load-bearing structure. External walls are of cavity construction dense concrete external slabs (4ft. wide by 1ft. 4in. high), and the internal a leaf of foamed slag blocks of a similar size.
Source: *PWBS No. 25.*

Figure 2 The Braithwaite House: the structural frame is of prefabricated steel sections, cold-rolled from light-gauge strip, which have welded joints. External cladding was normally ribbed asbestos-cement sheeting, with an internal dry lining. Source: *PWBS No. 23.*

Non-traditional houses thus continued to be viewed, as they had in the inter-war period, as 'makeshift' or 'temporary', where local authorities saw them as 'ten year' expedients after which the government would 'take the damned things away'.[53] Indeed misunderstanding, apparently, was endemic: 'much nonsense' being 'talked about prefabrication, chiefly because the meaning of the word has not always been fully understood'.[54] Did it mean, as was commonly implied, 'the manufacture of the complete house and its rapid and mechanical erection on site', or rather, as the Burt Committee and many architectural opinion-formers would rather have it, 'the application of certain factory methods in the mass production of certain component parts, thereby reducing site costs' for everyone?[55] Apologists for traditional methods again blamed the 'confusion of thought' on the 'political propagandists' of prefabrication:

> They insist on representing it as a technique providing a ready-made solution to the housing problem, gifted by progressive scientific thinkers to a grateful nation, but rejected by a backward and reactionary building industry, intent on preserving intact the twin citadels of craft monopoly and swollen profit. The public has been led to suspect that houses might now be rolling off the production line like typewriters or motor cars, were it not for the hidden and frustrating hand of vested interest.[56]

Contextually it is easy to see why 'traditionalists' felt themselves to be under siege, and prefabrication discourse thus privileged. The war years had heavily distorted the building industry. Labour productivity fell markedly, and remained stubbornly below pre-war levels after 1945. As demand for housing soared, one consequence was that the industry was viewed by politicians, economists and by scientific advisers alike as 'backward' and in crisis:[57] technically deficient, conservative in its use of mechanical plant, poorly trained, ignorant of recent research and resistant to the use of 'new methods'.[58] Bevan, as the minister responsible for housing, thought it 'vital' to provide an external 'stimulus to the building industry from outside ... by introducing new techniques'. Both he and Morrison, then in charge of co-ordinating domestic policy, understood permanent prefabrication in terms of a 'modern' alternative providing 'competition for construction'.[59] Cautiously stressing what had 'been tested and found by experience to be good' acquired an instinctive dissonance against this rhetoric.[60]

During the war the only development licences issued were for prefabricated or other experimental methods of building. Effectively this became the only game in town. Thus, the architect Richard Sheppard, surveying wartime developments, could justifiably comment in glowing terms on the much publicized proliferation of new prefabricated designs.[61] Sheppard favoured collaborative design, bringing together architect, production engineer, planner and structural engineer, complete with an ongoing scientific investigation into building materials and performance.[62] Indeed, the aggressive cultural promotion of new methods by certain architects and builders, the technocratic faith in experts, and the advance of scientific investigation, walked hand in hand.[63] The 'normal method of evolution of new forms of building' was a 'trial and error' process, taking considerable time. But time, it was argued, was now simply not available.[64] Science and technology offered an alternative path.

Yet the limits that the architectural modern movement and scientific influence had in redefining the house can be measured by looking briefly at what was actually constructed. The majority of systems houses built after 1945 were concrete-based (either cast *in situ* or prefabricated); second in popularity were those with a steel frame. Two builders (Wimpey and Laing) between them constructed a third of the 271,000 non-traditionals built in England and Wales in the ten years after the war: and both house types were cast *in situ*.[65] There was nothing new about such systems. Some 2,000 of Laing's 'Easiform' houses had been constructed before 1939, and slightly fewer 'no fines' (although not by Wimpey).[66] Both Laing and Wimpey were major speculative builders, initially producing non-traditionals because as 'normal building was expected to be limited by lack of traditional resources, it offered the prospect of a market'.[67] Neither house when assessed by the Burt Committee in 1944 was thought exceptional or worthy of praise. Instead they were judged 'a satisfactory alternative, if properly used, to brickwork', but 'not technically out-standing in such a way as to deserve special treatment'. Much greater interest, for example, was shown in steel-framed factory-manufactured structures. It was the latter that was preferred developmentally, for it was considered that the 'steel frame might form the basis of a true shop-fabricated demountable house'.[68] This belief that the output of the building industry could 'only be increased by using new [factory] methods' was deeply ingrained within the political structure also, so that early setbacks were discounted. Indeed, considerable disappointment was expressed that too many of the systems being introduced 'did not comply with the principles of prefabrication', and 'represented merely a method of construction'.[69] All systems construction was justified by the operating premise of its greater efficient use of scarce site labour. But the power of the ideal – that is shop-based prefabrication – flagged a heavy discrepancy between what was desired and what was finally produced.

Yet the broader idea of 'new methods' was certainly pinned, if not driven home, by the rapid expansion of systems construction. Although

Figure 3 Wimpey No-Fines House: The walls are of no-fines concrete, poured between wire mesh in a large framework. The first floor and roof was carried on steel beams, supported by stanchions.
Source: *PWBS No. 25.*

later falling, by 1948 non-traditional construction accounted for one third of all public-sector completions, a figure not subsequently exceeded until the high-rise boom of the late 1960s.[70] Indeed at one stage the Cabinet considered that some 75 per cent of permanent homes would be constructed by such 'labour saving methods'.[71] Nevertheless, even when successful, the governing philosophy of systems construction remained ambiguous – that is in terms of closure and stabilization through time. The point is again well illustrated by looking at cast *in situ* housing. The 'no-fines' and Easiform systems remained popular well into the 1960s and 1970s. In these years Wimpey alone was casting on average some 8,000 units per annum in England and Wales. 'No fines' options proved to be equally popular in Scotland. By and large such houses performed well: 'no fines' concrete walls were 'quickly built with very little skilled labour'.[72] Thus it was well liked by local authorities. But this popularity also rested with the appearance of 'no fines' construction, because once rendered externally, it passed for a traditionally built brick house. As Mass Observation observed, 'the less a prefab. looks like a prefab. the more people like it'. Some local authorities simply switched from brick to 'no fines' *in situ* concrete and back again, according to the availability of materials and labour. For others it became the system of choice.[73]

INNOVATION, EVALUATION AND LOTS OF PREJUDICE

Bowley argues that 'most of the systems proposed, and still more those actually used, were fundamentally similar to those used in the inter-war period'. New methods were not new at all. For her, perhaps the 'greatest innovation' was 'less in the actual methods and materials used, than in the development of the scientific assessment of performance.'[74] This offered the first objective technical standards against which performance could be measured (for example, thermal and sound insulation, resistance to fire and moisture penetration, stability, etc.). But 'neutral' state-sponsored science also became an evaluating weapon through which non-traditional methodologies could be and were promoted. Even before a detailed re-evaluation had commenced, science – in the form of the Building Research Station – spoke authoritatively about why earlier building systems had failed, and about its own current predictive powers to prevent such a physical reoccurrence.[75] At the same time the BRS readily admitted to the paucity of information existing on costs and labour content for housing generally – 'a remarkable dearth' of real systematic data: a major shortfall given that 'cost or demand on human effort is one of the final criteria in deciding on the merit of a new or modified form of construction.'[76] Science was to fill such voids in 'traditional knowledge': that which had served the industry 'so well in the past', but was 'no longer sufficient to meet the needs of modern conditions'.[77]

Thus, the BRS's publicly declared advice to architects was that, given the 'large fund of scientific knowledge of the physical, mechanical and chemical properties of building materials the problem of the design of new methods of construction presented no formidable obstacles.'[78] Such advice was disingenuous. Privately the BRS was advising the Burt Committee, for example, that the performance of light steel-framed houses was likely to be unpredictable after 25 years. It similarly refused to vouch for the efficiency of jointing systems utilized unless they could be tested in advance over time. Nevertheless, it was argued that the development of such houses for 'immediate post-war use' should not be 'prejudiced by certainties as to the ultimate length of life arising from either the corrosive or the jointing issue.'[79] Nor were jointing problems unique to steel houses. Similar shortcomings were encountered with pre-cast concrete houses but were again controversially gilded over in favour of production immediately.[80]

In re-evaluating non-traditional performance in the early 1950s, ministry officials referred to an immediate negative 'legacy of the past'. Poor technical performance, 'causing considerable trouble and expense in remedial measures', was one such failing, as were 'slow completion' rates, extra costs and the widely held belief that 'non-traditional houses were uglier that brick houses'.[81] In part this ignored the overall realities of the non-traditional versus traditional debate. Delays were endemic at that time across the board. Similarly, while non-traditional designs might 'accentuate' some of the difficulties of poor estate layout, and were certainly thought of as 'all looking alike', this was not a problem intrinsic or unique to them, but common on traditionally constructed estates too, 'where all individuality and homeliness have been lost in endless rows of identical semi-detached houses.'[82] Nevertheless, such an overview offered an accurate measure of perception: as one Regional Housing Officer remarked, there was an 'array of snipers abroad', all 'quick to seize upon and criticize any fault' in the non-traditional housing types.[83] As Bevan acknowledged, he expected positive steps to be taken to 'encourage' public authorities to place orders for non-traditional housing and 'to do everything possible to overcome the prejudice that still exists in all quarters against new building methods.'[84] That this was necessary reveals the large gap that existed between promoting government agencies and local authority consumers; one that remained after teething problems with earlier, more unreliable systems had been overcome.

It is useful also to contrast Bowley's post-operative view of the lack of technical innovation with that of contemporary advisers. Addressing building employers at the end of 1949, the Director of Research at the Ministry of Works began by noting that 'it had not been possible in the last thirty years for the building industry to provide a house at a rental within the means of the workers in the lower income groups.' Low-cost housing, he correctly summarized, 'has had to be subsidised'.[85] He then outlined the recent advances made, continually contrasting these against a slow-moving building industry 'firmly entrenched in traditional practice ... founded upon craftsmanship' and making materials fit on site. 'The new methods of production differ from the old', he stated, in that 'by definition, every stage is planned and the whole process is in every case rigorously controlled by the requirements of production'.

The actuating theory has been that structural components should be made to accurate dimensions in factories and they should then be assembled on site with a minimum of labour. In fact, the expected economies in labour have been achieved handsomely. The components have been made accurately and they have been assembled quickly and easily.[86]

Indeed, he concluded, 'the pity of it all' was that such techniques had so far been limited only to the shell of the house, as he announced plans to fabricate completely the interiors and finishings in the factory also.

There is no record of audience reaction to such claims, although it is unlikely to have been sympathetic because most contractors continued to favour traditional methods.[87] It would have been less sympathetic still had the audience fully appreciated how radically estimates of man-hour savings then being presented to them had fallen from the initial figures that had first underpinned prefabrication policy. As Bernal, then chairing the MOW Scientific Advisory Group, pointed out in early 1946: 'It appears that much of the time-saving in certain types of prefabrication is more apparent than real.'[88] Again it was 'outstanding' houses like the Reema and Wates that were praised by the BRS: those predominately prefabricated off-site and thus most readily conforming to the ideal of factory methodology. The government's report on non-traditional performance reminded its audience that success 'depended on the basic principle of exploiting the machine to the full, assembling very rapidly on the site a number of accurately dimensioned units which require no elaborate fitting together.'[89] New methods, however, were still no cheaper than traditional forms of construction, and indeed only 'no fines' offered immediate prime-cost savings.[90] Nevertheless, the development of pre-cast concrete housing systems was to be dominated by a single theme: bigger panels incorporating an increasing number of functions (for example, a pre-finished internal skin). The Reema and Wates systems were early exemplars. As the BRS noted, it seemed reasonable 'to make the units as large as the mechanical equipment can conveniently handle'.[91] Here we see a glimpse of the future. Building high, the next stage, was always to be more expensive. This led to the lowering of specifications and design standards in the utilitarian blocks of the 1960s and 1970s. Practical limits existed also to the economic size of off-site prefabricated units, and high factory-based overhead costs married uneasily with fluctuating demand. Moreover, the efficiency claims made for systems construction as likely reflected more efficient site management structures than the system methodology *per se*.[92]

Such ante-dated realities sit uneasily with science's predictive confidence in the 1940s. And perhaps post-knowledge criticism is unfair. Well, not really. The authority of such discourse was itself created initially by contrasting an earlier lack of knowledge with what was now 'understood' about material behaviour and areas such as works study. Contemporaries also already suspected that productivity gains attributed to non-traditionals were a product of better site organization. Indeed the concluding remarks of the first Burt Report of 1944 captured exactly this

Figure 4 The Wates House: The walls are built of pre-cast reinforced concrete units of a tray shape (7ft. 6in. high by 2ft., 3ft. or 4ft. wide). The internal wall lining is formed of factory-made timber-framed units faced with plasterboard. Source: *PWSB No. 25.*

latter sentiment.[93] When such questions were raised, however, the answer, although present in the analysis, was not one that was emphasized or headlined. It remained buried, neutrally, in the main body text.[94] We can contrast this 'sleight of hand' with the emphasis taken by those major overarching enquiries into building efficiency at that time. These virtually discounted the probability of gains coming from shell prefabrication or indeed from prefabrication generally.[95] Nevertheless, for politicians, science's role was axial; its course and agenda set. Writing to Morrison, the Minster of Works put this perfectly. All agreed on the 'importance of research work in the field of building and civil engineering from the point of view of the efficiency', and making 'the most economical use of man-power and material as well as from the point of view of cost.' The 'time has come when the underlying problem of modernising the building industry and putting pressure on it to become technically efficient will have to be faced.'[96] Indeed, construction's very backwardness was publicly defined in terms of its reluctance, as an 'ancient industry, rooted in tradition and craft practice', to adopt mechanization and prefabrication – the products of scientific investigation.[97] If we factor in the prostituted use of this broader scientific discourse through systems advertising and 'neutral' technical reviews (where such language underlined manufacturers'/ designers' claims), the rhetoric of science had a significant impact in promoting systems mentalities.[98] But it also exacerbated the gap between the 'new' and 'old', between traditional and non-traditional. Thus, only in one sense did it aid understanding, and then only partially.

Bowley, writing at the time of the later high-rise boom, acknowledges that, while most of the non-traditional low-rise systems used after 1945 were 'fundamentally similar' to those used between the wars, there were what she terms 'novelty' innovations. Included here was the use of the large pre-cast load-bearing panel (that is the Wates and Reema systems already noted), and of lightweight steel-frame fabrication. Most of the innovative steel-frame solutions, she notes, failed to be developed – partly because government policy towards steel houses changed significantly due to post-war steel shortages. Other 'genuine innovations' could be found in the use of aluminium and plywood for the construction of internal and external skins.[99] Some historians have been more generous. Finnimore, for example, points out that the majority of non-traditional dwellings 'used materials new to housing such as concrete, steel and laminates', noting again particularly those 'systems which used the latest techniques in light weight steel fabrication'. He concludes that of 'more impact than the real extent of mechanised production was the innovative nature of the materials and methods of construction used'.[100] Such praise requires obvious qualification because steel and concrete were central components of earlier inter-war systems. Even those types noted as innovatory in their use of materials – like the BISF type, of which over 31,000 were built – were modelled loosely on earlier inter-war types like the Dorlonco (itself then considered to be an 'important technical step forward').[101] For some house types, such as the steel prototype Braithwaite, it was the method of 'clip' jointing that was deemed to be important.[102] Indeed a frequently stated reason for the post-war 'failure' – if such it was – of non-traditional housing was that it focused on too many 'novelty' individual private designs, so markets for each were correspondingly limited, when what was really needed was a broader trajectory of prefabrication and standardization.[103]

There is some sense to this. Certainly the production of certain house types was driven forward by little but entrepreneurial enthusiasm, and this was really part of the infectious post-war hype associated with non-traditional methods. The Woolaway – a pre-cast system – was one such: poorly reviewed in terms of its physical and design attributes, and lacking the necessary finance. Even the minister responsible recognized its limited potential: 'only one of a hundred' such approved systems, and 'not of any outstanding merit', but nevertheless encouraged as 'a useful supplement to traditional forms'. Only some 4,300 were eventually built before the

Figure 5   The British Iron and Steel Federation (BISF) House: as illustrated here, the frame is of light-gauge sections, cold formed from mild-steel strip and fabricated by spot and ridge welding. Cavity brickwork/blockwork extended to first floor height, with ribbed steel sheeting and internal lining above.
Source: *PWBS No. 23.*

Figure 6   The Woolaway House. The walls are of post-and-panel pre-cast concrete construction. The posts were full single-storey height, the panels are of aerated concrete, normally 4ft. high by 2 ft. wide, and rendered *in situ* externally.
Source: *PWBS No. 25.*

company went bankrupt, and that after the government had already offered surety and actively manipulated the housing market to guarantee orders.[104] Indeed, it is noteworthy that even though extra financial subsidies were terminated by 1948, government agencies actively continued to manipulate the public-housing market to support non-traditional producers.[105] Nevertheless, certain types were popular; or, at the very least, less unpopular. It was the larger, national companies that had fewer difficulties filling order books (again notably Wimpey and Laing, but to a lesser degree Wates), while the smaller, locally orientated producers like Reema and Spooner struggled to obtain a continuity of work, raising unit costs because of high central overheads.[106] There were exceptions. The Cornish Unit, of which over 22,000 were built in the decade after 1945, proved popular in the south-west, partly because it was enthusiastically promoted by regional staff, partly because of the shortage of traditional builders in that area, but also because of its attractive, distinctive design: indeed Bevan wanted one as a country retreat.[107]

CONCLUSION

Customers (householders and local authorities) were noticeably less resistant to certain non-traditional designs after 1945 than they had been before 1939: meaning that how non-traditionals were understood and contextualized had changed. Thus, there was a greater acceptance of certain house types that previously had been shunned. Nevertheless, limits still existed as to what was more and was less acceptable, measured against the cultural 'gold standard' of traditional practice. This 'marked preference for the familiar in housing, for "quality", whether real or imagined', meant, according to one housing historian, that the foremost '"problem" [facing non-traditional diffusion] was social and political, for unless all sections of the community could agree on the desirability of mass-produced houses as consumer goods, there could be no success'.[108] Two points spring from this. The first is an assumption that technological 'meaning' and 'direction' had also been satisfactorily resolved. The second revolves around the question of 'consensus', as outlined above or in terms of an 'end to controversy'. How should we measure it, how do we mark its attainment, is it a useful tool?

Clearly there are quantitative and qualitative problems with both points. With hindsight, we can judge that future trends in systems construction were to follow the paths favoured through late-1940s and early-1950s low-rise design. That is, preference would be given to large prefabricated concrete panels and, at the other extreme, cast *in situ* concrete housing. The former was strongly advocated by scientific, technological and modernist architectural opinion, and was explicit in the 'ideal' promoted by politicians (the paymasters of public housing) – that modernizing a 'backward' building industry involved the transfer of production from site to factory. *In situ* construction, by contrast, fell outside this rubric. Yet, ironically, it was here that customer resistance to changing consumption patterns was noticeably weaker. Indeed, here we find positive acceptance – or certainly significantly less controversy amongst the relevant social groups. 'No fines' construction in practice meant building a 'traditional' house using non-traditional means, for the same or less money, more quickly, and with fewer problems in terms of labour and materials shortages. Off-site prefabrication, however, had a different meaning. Initially, it represented standardization and the manufacture of component parts for the industry as a whole. But in terms of popular exposure – which itself imparted important meaning – it signified whole-house manufacture, and in terms of the implicit promises made: the mass production of houses. This, it seems, was the rehearsed 'dream' of publicists, manufacturers and modern architects.[109]

This new ideal was widely misunderstood, and indeed widely misrepresented. It is significant that even at times of acute housing shortages, customers at best subscribed to it only through a 'better than nothing' mentality. It was bitterly resisted, too, by the traditional industry – which rightly viewed it as a negative commentary on its own abilities – and by many architects. Was this unimportant? A consensus promoting prefabrication was unnecessary, it might be argued, because of the social authority of the groups supporting the proposition: those thought capable of legitimately speaking for the future (scientists, technologists, ideo-logues). Consensus, anyway, is an ideal construct: a widespread willingness to agree through common understanding across social groups is in fact unlikely. Perhaps then a better question to ask is to what degree, comparatively, was agreement reached – this being a less exacting requirement? Even here, however, there was no end to controversy, no social closure. The quality of conversation between consumer and producers remained, for the most part, marginal, and certainly not informing. So solutions were imposed, rather than agreed. No longer was successful innovation to be measured by an ability to mass-produce houses, but, instead, the industrialization of building simply meant that non-traditionals should be able to compete economically, using less skilled site labour.[110] And, even against these more limited criteria, unanimity over prefabrication's achievements was lacking. This is not to say that a future course of direction had not been established. The largely unquestioned authority of science and technology, juxtaposed against continuing negative perceptions of traditional construction practice, continued to set the rules. That this failed to achieve an overarching hegemonic control in the ten years after 1945 because of consumer resistance does not mean that this did not occur later. But even then customers remained isolated, ignored; closure and stabilization thwarted. The results were the inadequate, inappropriate solutions of the 1960s and 1970s.

Notes and References

1. Public Records Office (hereafter PRO), HLG 36/18, Report of the Design of Dwellings Sub-Committee, 8 February 1944, para. 116.

2. Raphael Samuels, *Island Stories: Unravelling Britain. Theatres of Memory, Volume Two* (London, 1998), 101–24.

3. Christopher Powell, *The British Building Industry Since 1800: An Economic History* (London, 1996), 134.

4. Nick Hayes, 'Making Homes by Machine: Images, Ideas and Myths in the Diffusion of Non-Traditional Housing in Britain 1942–54', *Twentieth Century British History*, 1999, 10: 295.

5. 'Prefabrication: What will it mean in housing', *Picture Post*, 4 March 1944.

6. 'Introduction: Strategic Research Sites', in Weibe Bijker, Thomas Hughes and Trevor Pinch (eds.), *The Social Construction of Technological Systems; New Directions in the Sociology and History of Technology* (Cambridge, Massachusetts, 1987), 191.

7. Mass Observation, *An Enquiry into People's Homes* (London, 1943), 53, 156, 21; Hayes, *op. cit.*(4), 297–8.

8. Steven Fielding, Peter Thompson and Nick Tiratsoo, *'England Arise!' The Labour Party and Popular Politics in 1940s Britain* (Manchester, 1995), 36–7.

9. Weibe E. Bijker, 'The Social Construction of Bakelite: Towards a Theory of Invention', in Bijker *et al.*, *op. cit.* (6), 171–4.

10. Steven Groák, *The Idea of Building: Thought and Action in the Design and Production of Building* (London, 1992), 38–9.

11. 'Introduction', Biker, *op. cit.* (6), 12–13.

12. Marian Bowley, *The British Building Industry: Four Studies in Response and Resistance to Change* (London, 1966), 325.

13. *Ibid.*; Steven Groák and Graham Ive, 'Economics and Technological Change: Some Implications for the Study of the Building Industry', *Habitat International*, 1986, 10: 116–18, 125–8.

14. David Mowery and Nathan Rosenberg, *Technology and the Pursuit of Economic Growth* (Cambridge, 1989), 9; Keith Pavitt, 'Sectoral patterns of technological change; Towards a taxonomy and a theory', *Research Policy*, 1984, 13: 365.

15. W.J. McGhie, 'The Industrialisation of the Production of Building Elements and Components', *Production of the Built Environment*, 1983, 4: 32, 34–5; Groák, *op. cit.* (10), 142; D. Sugden, 'The Place of Construction in the Economy', in D.A. Turin (ed.), *Aspects of the Economics of Construction* (London, 1975), 16–17.

16. Bengt-Åke Lundvall, 'Innovation as an interactive process: from user-producer interaction to the national system of innovation', in Giovanni Dosi, Christopher Freeman, Richard Nelson, Gerald Silverberg and Luc Soete (eds.), *Technical Change and Economic Theory* (London, 1988), 349–58.

17. Pavitt, *op. cit.* (14), 356; McGhie, *op. cit.* (15), 35.

18. Nathan Rosenberg, *Inside the black box: Technology and Economics* (Cambridge, 1982), 66; Bowley, *op. cit.* (12), *passim*.

19. PRO DSIR 4/592, I. Bowen, 'Productivity in the Building Industry'; C.M. Kohen, *Works and Buildings* (London, 1952), 448.

20. Groák, *op. cit.* (10), ch. 3; Tavistock Institute, *Interdependence and Uncertainty: A Study of the Building Industry* (London, 1966), 39, 57.

21. Bowley, *op. cit.* (12), 350.

22. *Ibid.*, 207–21.

23. Patrick Dunleavy, *The Politics of Mass Housing in Britain 1945-1975: A Study of Corporate Power, and Professional Influence in the Welfare State* (Oxford, 1981); John Gold, *The Experience of Modernism: Modern Architects and the Future City 1928-1953* (London, 1997), ch. 7; Fielding *et al.*, *op. cit.* (8), 110–12; Junich Hasegawa, 'The Rise and Fall of Radical Reconstruction in 1940s Britain', *Twentieth Century British History*, 1999, 10: 137–61.

24. Dunleavy, *op. cit.* (23), 3, 59-60, 342-52.

25. Brian Finnimore, *Houses From the Factory: Systems Building and the Welfare State* (London, 1989), 240–1.

26. Richard Nelson and Sidney Winter, 'In Search of Useful Innovation Theory', *Research Policy*, 1977, 6: 56–60; Giovanni Dosi, 'The Nature of the Innovation Process', in Dosi *et al.*, *op. cit.*, (16), 222, 228.

27. Finnimore, *op. cit.* (25); Dunleavy, *op. cit.* (23); Barry Russell, *Building Systems, Industrialization, and Architecture* (London, 1981); Robert McCutcheon, 'Modern Construction Technology in Low-Income Housing Policy', *History of Technology*, 1990, 12:136–76.

28. Dunleavy, *op. cit.* (23), 177–8, 348–50; Finnimore, *op. cit.* (25), 18, 44–5, 238–45;

Oonagh Gay, 'Prefabs: A Study in Policy-Making', *Public Administration*, 1987, 65: 407–22; McCutcheon, *op. cit.* (27).

29. Henkvan den Belt and Arip Rip, 'The Nelson-Winter-Dosi Model and Synthetic Dye Chemistry', in Bijker *et al.*, *op. cit.* (6), 142.

30. Giovanni Dosi and Luigi Orsenigo, 'Coordination and transformation: an overview of structures, behaviours and change in evolutionary environments', in Dosi *et al.*, *op. cit.* (16), 18.

31. Finnimore, *op. cit.* (25), 18, 245; Hayes, *op. cit.* (4).

32. Hayes, *op. cit.* (4); Robert McCutcheon, 'Major Participants in the UK Building Industry, 1964–1977', *Habitat International*, 1988, 12: 105–16.

33. Hayes, *op. cit.* (4).

34. Trevor Pinch and Wiebe Bijker, 'The Social Construction of Facts and Artifacts: Or How the Sociology of Science and the Sociology of Technology Might Benefit Each Other', in Bijker *et al.*, *op. cit.* (6), 44.

35. *National Builder*, March 1944. For the full report, see Ministry of Works, *Methods of Building in the USA: The Report of a Mission appointed by the Minister of Works* (London, 1944). In its conclusions it makes hardly a mention of prefabrication.

36. J.B. Priestley, *The English Journey* (London, 1934, republished London, 1984), 300–3.

37. David Jeremiah, *Architecture and Design for the Family in Britain, 1900–70* (Manchester, 2000), 132; Brenda Vale, *Prefabs: A History of the UK Temporary Housing Programme* (London, 1995), ch. 3; Finnimore, *op. cit.* (25), 41–3.

38. 'Fifty Thousand Brides Envy Her', *Picture Post*, 27 October 1945; 'A House Goes Up in a Day', *Picture Post*, 21 September 1946; 'Prefabrication: What Will it Mean in Housing', *Picture Post*, 4 March 1944.

39. Richard Sheppard, 'US Wartime Housing', *Architectural Review*, August 1944.

40. Correlli Barnett, *The Audit of War: The Illusion and Reality of Britain as a Great Nation* (London, 1986).

41. PRO HLG 94/7, Letter from McGuiness (Secretary of State for Scotland) to Burt Committee, 2 July 1943.

42. Mass Observation, File Report 2270B, 'First Report on "The Post-War Home Exhibition"', 28 July 1945 (Harvester Microfiche, 1983).

43. *Ibid.*

44. Hayes, *op. cit.* (4), 291–3; David Dean, *The Thirties: Recalling the English Architectural Scene* (London, 1983), 9.

45. 'RIBA on House Production', *Architects' Journal*, 19 April 1945; 'Prefabrication and the Builder', *Architects' Journal*, 6 December 1945.

46. 'Prefabrication: A Policy', *ibid.*, 20 December 1945.

47. PRO HLG 94/1, Ledeboer, Burt Comm mins, 27 March 1943.

48. R. Fitzmaurice, 'Scientific Research on Alternative Methods of Construction for Permanent Houses: Part 1', *Journal of the Royal Institute of British Architects* (hereafter *RIBA Journal*), April 1947.

49. Post-War Building Studies (hereafter *PWBS*) *No. 1, House Construction: by an Interdepartmental Committee Appointed by the Minister of Health, Secretary of State for Scotland and the Minister of Works* (The Burt Committee), (London, 1944), 2–3.

50. *Architects' Journal*, 6 April 1944. R.B. White, *Prefabrication: A History of its Development in Great Britain* (London, 1965), 154, makes a similar point.

51. John Wilton, 'Prefabrication: Problems in the Development of a Technique', *National Builder*, July 1946.

52. Mass Observation, *loc. cit.* (42), File 2270B; Mass Observation, File 2360, 'Modern Home Exhibition', 28 March 1946.

53. *Ibid.*; Dex Harrison, 'An Outline of Prefabrication', in John Madge (ed.), *Tomorrow's Houses: New Building Methods, Structures and Materials* (London, 1946), 118; Richard Sheppard, 'Developments in Post-War Housing in 1944', *Architects' Journal*, 18 January 1944.

54. 'Prefabrication: A Definition', *Architects' Journal*, 13 December 1945; Hugh Antony, *Houses: Permanence and Prefabrication* (London, 1945), 42.

55. *PWBS No. 1*, 1; 'Prefabrication and the Builder', *Architects' Journal*, 6 December 1945; 'Prefabrication II: A Policy', *Architects' Journal*, 20 December 1945.

56. John Wilton, 'Prefabrication: The Need for Rational Development', *National Builder*, October 1946.

57. Nick Hayes, 'Forcing Modernisation on the "one remaining really backward industry": British Construction and the Politics of Progress and Ambiguous Assessment', *Journal of European Economic History* 2002, 31: 321–50.

58. For example, PRO CAB 124/553, Report by Advisory Council on Scientific Policy, 27 June 1947; CAB 124/553, Lord President's mins, 9 July 1948, 13 May 1949.

59. PRO CAB 134/642, Bevan, 'House Building Costs', 16 June 1949; CAB 124/474, Morrison to Maud, 1 August 1945.

60. H. Scott-Hume, 'Organisation for Production', *National Builder*, January 1945.

61. 'Developments in Post-War Housing in 1944', *Architects' Journal*, 18 January 1945.

62. 'Prefabrication', *RIBA Journal*, May 1944.

63. Hayes, *op. cit.* (4); John Stevenson, 'Planners' Moon? The Second World War and the Planning Movement', in Harold Smith (ed.), *War and Social Change: British Society in the Second World War* (Manchester, 1986), 58–77.

64. Fitzmaurice, *op. cit.* (48); Richard Sheppard, *Prefabrication in Building* (London, 1946), 12, 15–17.

65. Ministry of Health, *Housing Returns for England and Wales 1946–55* (London, date).

66. In 'no fines' concrete, sand was omitted, lightening and cheapening the mix, and thus lightening, too, the shuttering needed. This concrete mix also improved thermal insulation and prevented capillary attraction through the wall.

67. Bowley, *op. cit.* (12), 213.

68. *PWBS No. 1*, 43-47, 53–56; PRO HLG 101/55, Procter to Taylor, 12 December 1944; HLG 94/5, 'First Interim-Report of the Inter-Departmental Committee', October 1943; HLG 94/7, Burt Comm. mins 17 June 1943, BC 46 'Experiments in House Construction'; HLG 94/6, Burt Comm. mins 15 April 1943, BC 36 'Report on Steel Framed Houses'.

69. PRO CAB 134/642, mins 16 June 1949; CAB 87/36, mins 29 August 1945.

70. Finnimore, *op. cit.* (25), 262.

71. PRO CAB 87/37, H (45) 13, 9 March 1945, note by Sandys.

72. Finnimore, *op. cit.* (25), 268–71; White, *op. cit.* (50), 191–2; *Housing Returns for Scotland 1946–55* (London, date); *PWBS No. 25, House Construction: Third Report of Inter-Departmental Committee* (London, 1948), 37.

73. Bowley, *op. cit.* (12), 224; Mass Observation, File 2360, *loc. cit.* (42); Finnimore, *op. cit.* (25), 51–55; Nick Hayes, *Consensus and Controversy: City Politics in Nottingham 1945–1966* (Liverpool, 1996), 42–3, 56, 81. The clear exception, in terms of appearance and popularity, were the temporary house types. For why this was so, see Vale, *op. cit.* (37), ch. 1.

74. Bowley, *op. cit.* (12), 228, 231.

75. PRO HLG 94/1, Building Research Station, Note No. 844, February 1943.

76. PRO HLG 94/4, Boxall to Symon, 3 July 1943; HLG 94/7, Burt Comm. mins, 10 August 1943, BRS, 'Memo on Experimental Building'.

77. Ministry of Works, *Advisory Council on Building and Research and Development: First Report* (London, 1949), 12.

78. Fitzmaurice, *op. cit.* (48).

79. PRO HLG 94/8, Burt Comm. mins, 13 July 1944 and 17 August 1944.

80. *Ibid.*, 29 November 1945 and 17 January 1946. See Russell, *op. cit.* (27), 651–68, for subsequent problems generally.

81. PRO HLG 101/635, 'Expansion of Non-Traditional Programme', 13 May 1952.

82. PRO HLG 36/21, CHAC PW 64, 'The Appearance of Housing Estates', 1947; Mass Observation, File 2360, *loc. cit.*

83. PRO HLG 102/371, Smith to Hickinbotham, 7 January 1952.

84. PRO CAB 134/642, 'House Building Costs', 16 June 1949.

85. R. Fitzmaurice, 'Changes in Building Technique', *National Builder*, December 1949.

86. *Ibid.*

87. See, for example, *National Federation of Building Employers Annual Report 1943*, 10; *National Builder*, March 1945.

88. Hayes, *op. cit.* (4), 303; J.D. Bernal, 'Building Research', *National Builder*, March 1946.

89. Ministry of Works, *National Building Studies Special Report No. 4: New Methods of House Construction* (London, 1948), 4; PRO HLG 101/371, Hickinbotham to Wilkinson, 4 February

1952; HLG 101/371, Policy Comm. on Non-Traditionals, c. Jan 1952.

90. Ministry of Works, *National Building Studies Special Report No. 10: New Methods of House Construction Second Report* (London, 1949); Bowley, *op. cit.* (12), 243–8; Hayes, *op. cit.* (4), 304–7.

91. R. Fitzmaurice, 'Scientific Research on Alternative Methods of Construction for Permanent Houses Part 2', *RIBA Journal*, May 1947.

92. Finnimore, *op. cit.* (25), 169–83, 197–201; Dunleavy, *op. cit.* (23), 84–8; Graham Towers, *Shelter is Not Enough: Transforming multi-storey housing* (Bristol, 2000), 52–62; Powell, *op. cit.* (3), 187–9.

93. *PWBS No. 1*, 139.

94. J. Bronowski, 'Operational and Statistical Research in Building', *Architects' Journal*, 29 March 1951; *NBS No. 4*, *op. cit.* (89), 3, *passim*.

95. Anglo-American Council on Productivity, *Building: Report on a visit to the USA in 1949 of a Productivity Team representing the Building Industry* (London, 1950), 40; Ministry of Works, *Working Party Report Building* (London, 1950), 39; Ministry of Health, *The Cost of House Building: First Report of the Committee of Inquiry* (London, 1948), 49–50.

96. PRO CAB 124/553, Tomlinson to Morrison, 7 January 1946; Morrison to Key, June 1947.

97. R. Fitzmaurice, 'Influence of Mechanisation and Prefabrication on Techniques and Cost of Building', *National Builder*, October 1951; Durbin, *Operative Builder*, September 1948.

98. Dunleavy, *op. cit.* (23), 110–114; Hayes, *op. cit.* (4), 288–91; McCutcheon, *op. cit.* (32), 113.

99. Bowley, *op. cit.* (12), 228–31.

100. Finnimore, *op. cit.* (25), 44.

101. Sheppard, *op. cit.* (64), 48; James Strike, *Construction into Design: The Influence of New Methods of Construction on Architectural Design 1690–1990* (Oxford, 1991), 123.

102. Sheppard, *op. cit.* (64), 56, 59; PRO HLG 101/56, mins 13 February 1946; *Architects' Journal*, 5 October 1944.

103. White, *op. cit.* (50), 151–4, 160, 300; Strike, *op. cit.* (101), 154.

104. PRO HLG 101/56, Bevan, note c. Oct 1945; Scott to Symon, 19 October 1945; *PWBS No. 25*, 29–33.

105. Finnimore, *op. cit.* (25), 63–6; Hayes, *op. cit.* (4), 304–7.

106. PRO HLG 101/635, 'Expansion of the Non-Traditional Programme', 13 May 1952; Hickinbotham to Wilkinson, 21 May 1952.

107. Hayes, *op. cit.* (4); Michael Foot, *Aneurin Bevan 1945–1960* (London, 1973), 81.

108. White, *op. cit.* (50), 127–8.

109. For example, Le Corbusier, *Towards a New Architecture* (Paris, 1923, reprinted Woburn, 1970 edn.), 229–65; Walter Gropius, *Bauhausbücher, Vol. 3: Ein Versuchshaus des Bauhauses* (Munich, 1924), reprinted in Gropius, *Scope of Total Architecture* (London, 1956), 143–54; Sheppard, *op. cit.* (64); Antony, *op. cit.* (54).

110. McCutcheon, *op. cit.* (27).

Impacts of Technology Reassessed: A Retrospective Analysis of Computing Technology

JAN VAN DEN ENDE

INTRODUCTION

Does technology have impacts? According to popular opinions it clearly has. Many people even attribute the most important changes in society to the development of technology. For scholars in the field of technology and innovation studies the answer is not so easy. Since the advent of social constructivism we have considered technology as created by specific humans at specific times and places.[1] The same holds for the social changes associated with the application of technology, so technology has no impacts of its own. An example of this forms the study *Does Technology Drive History?*.[2] The authors discussed the issue of the autonomy of technological development and the impact of technology on society. They hardly reached a clear conclusion except for Thomas Misa's proposal to accept that technological determinism would come to the fore on the macro level, and that social determinism would be most prevalent on the micro level, and that on the meso level the two conceptions interact.[3] It would mean that technology would only have impacts on the macro level, which is of course a rather unsatisfactory idea.

In this paper I will take up the question regarding the impacts of technology, making the issue time-and-technology specific. The question is in which fields and periods we can meaningfully speak of the impacts of technology. The background is that many of the discussions on the impacts of technology, such as in *Does Technology Drive History?*,[4] are 'categorical', in the sense that questions are posed concerning technology development in general, not concerning specific fields of technology and specific periods. I will argue that the degree to which impacts occur depends on two types of enabling factors for innovation: the degree of knowledge development and the degree of changes in demand. As a consequence, the impacts of technology development vary according to period and field of technology.

Referring to knowledge and demand reminds us of the debate in the field of technology studies in the 1970s and 1980s between protagonists of the technology-push approach and the demand-pull approach. The question was whether knowledge development or market demand was the most important driver of innovation. That debate remained largely unsettled and led to the conviction that the developments of technology and society are co-evolutionary, and that both technology-push and demand-pull are important for innovation.[5] One of the reasons was that in the discussion knowledge (technology) and demand referred to very different things: the motivations of innovators (the availability of new knowledge versus the objective to meet specific demands) or the source of innovation (university research versus practitioners' inventions). In this paper I analyse the development of knowledge and demand as enablers of innovation in specific fields of technology. In my view new technological knowledge, whether developed by innovators themselves or by specialized academic or industrial researchers, is an important resource for innovators in developing new products and is the most important determinant of the degree to which impacts occur.[6] That focus in this paper does not preclude the importance of the actors involved in the development of technology in a specific field for the generation of impacts. However, the individual or collective behaviour of actors and the character of the actor network explain only part of the degree to which impacts occur in a specific field of technology as a whole.

The claim of this paper is that new knowledge often facilitates improvements in price-performance ratios of the products on offer.[7] These improvements in turn form a stimulus for target users to adopt the technology in the first place, but also to change their activities in conformance with the possibilities of the new technology, and to induce others to do the same. As a consequence, in the case where knowledge development is the main enabler of innovation in a specific field, technology may create impacts. When demand, even latent demand, is the main enabler, and when there is no major new knowledge applied in innovation, technology is in fact adapted to the requirements of users or potential users. The price or performance of the technology may change (the price may also increase to compensate for a higher quality) but in this case overall no important changes in price-performance ratios will occur, and the impacts will in general be far more limited.

For the analysis of this paper I make a longitudinal analysis of the development and application of computing technology in the twentieth century. This was a field in which many new technologies were developed, which spread throughout society. I will demonstrate that different periods can be distinguished with respect to the enabling role of knowledge development and demands in this field of technology. For one of these periods (1900–60) I discuss some specific cases of the applications of innovations. The reason for investigating these cases is that the importance of demand can only be analysed by investigating adoption decisions of users in the course of time. For the period 1960–2000 an analysis of the changes in the price-performance ratios is the most important basis for the overall analysis.

In the paper I first briefly discuss the main trends in the historiography and the history of computing. Based on the empirical material I divide the history of this field in the twentieth century into three periods: 1900–60, 1960–90, and 1990–2000, arguing that in these three periods the dominant enabling factors behind innovation were different. The case studies that I discuss for the period 1900–60 are two computing practices in the Netherlands (computing in water works and statistical data processing). In the conclusion and discussion I reflect on the generalizability of these findings in other fields of technology and on their relevance for the history of technology.

THE HISTORY OF COMPUTING TECHNOLOGY

Until well into the 1980s the literature on the history of computing technology was very traditional in its explanation of the development and spread of new technologies, including its vision of their impacts. A traditional internalist stance prevailed, focusing on inventions and the people that made them, and explaining the growth of the field with reference to knowledge developments.[8] As far as social factors were discussed, these were often restricted to the role of the military in the development of early computers.[9] Since the 1990s the literature has paid more attention to the history of the computer industry and to the wider social contexts of the applications of computing technology.[10] The discipline increasingly followed the general tendency in technology studies to integrate different forces in the explanation of technological development, however, with the same consequences for the explanation of social impacts.

The abundant literature on the information society from the 1970s and 1980s has dedicated far more attention to societal impacts.[11] However, this literature has taken a similar stance as the traditional internalist history of computing literature as far as the explanation of technical change is concerned. Authors in this field often assumed that technological developments, which they considered to be mainly autonomous, are the main driving force behind societal changes, which consequently had the character of impacts.

An exception in both fields was James Beniger, who emphasized the societal background of the emergence of the computer revolution in his book *The Control Revolution*.[12] According to Beniger, the diffusion of computing technology in the twentieth century has to be explained with reference to a general control crisis, which emerged in Western societies in the nineteenth century, and which continued during the twentieth century. In his eyes the growing speed and complexity of the material processing system (encompassing production, distribution and sales systems) had caused the control crisis. Beniger's analysis is interesting since he considered societal changes on the macro level resulting from the material-processing system to be the main stimulus for the development of computing technology, which in turn generated a control revolution. To what extent this control revolution has to be considered an impact of the new technology remains unclear.

During the period from 1900 to 1960 the field of computing technology included three different types of activities: data processing, technical and scientific computing and computing in process control. The most important computing technologies applied in the field of data processing were desk calculating machines and punched-card machines, whereas engineers and scientists applied slide rules, graphical aids and analogue computing machines for technical and scientific computing activities. Electrical and mechanical control devices (supported by servomechanisms) were used for process control. Occasionally, technologies from one field were applied in another, such as punched-card machines that were used for technical scientific computations by some specialized computing bureaus in the US and UK.[13] In the 1950s, the digital computer united the three fields to a considerable extent.

In all three fields the volume of application of computing technologies grew in the period up to the digital computer. Scientists, engineers and manufacturing firms also developed and introduced many new computing technologies. They were applied in industrial and university research laboratories, which were newly established in this period, and which performed many computational tasks.[14] Government agencies and firms applied computing technologies for their administrative tasks.[15] And in industry computing devices were used for automatic control, which became an issue in the period around the Second World War.[16] Often different computing technologies were applied in parallel by the same organization.

Two cases of specific users of computing technologies, both from the Netherlands but which are also representative of many other users in other countries, provide insight into the background of the growing degree of application of computing technology, and its impacts in this period: tidal calculations and statistical data processing. The first case is an example of technical scientific calculations, the second one of data processing. The two cases have furthermore been selected because these were large-scale applications of computing technology in this period.

In the Netherlands, tidal calculations served to predict changes in the tidal pattern and water levels during storms, as part of a preparation for hydraulic works that served to reclaim land from the sea and to protect the land against floods.[17] These works involved the enclosure of the Zuiderzee by a 30 km dam, a project that was executed between 1918 and 1932, and the Delta Works, which were carried out between the mid-1950s and 1986. In the second half of the nineteenth century, civil engineers had applied intuitive methods to predict the effects of hydraulic works on tidal patterns and on the propagation of storm surges, and had made simple computational estimates. Between 1900 and 1960, engineers and scientists introduced four new methods for these computations: manual computational methods (1918), electrical analogue computing methods (1945), large hydraulic models (1948) and the digital computer (1956).

The Zuiderzee works was the first hydraulic work for which the old intuitive and simple computational methods were considered to lack the required accuracy and certainty. Civil engineers, local and regional authorities and members of parliament requested better computations. The government invoked the assistance of the then famous physicist H.A. Lorentz to solve the problem. He worked together with a civil engineer for eight years to produce exact calculations to predict the changes in water movements. Of these, they spent numerous months on manual computing work alone. Their work was a success: the actual increases in tidal movements as a consequence of the construction of the dam were very close to their predictions (errors being in the order of 10 per cent of the predicted change in tidal range). After this project the National Public Works Administration started practising the manual computing method on a much larger scale in the 1930s, when it appointed about ten human computers just to perform such calculations for another region, the Delta area in the south-west of the Netherlands, where the Rhine, Meuse and Scheldt flow into the North Sea. Experiments resulting in the second method, electrical analogue computing models, started in the mid-1940s, resulting in the construction of two large analogue computers for this problem by the National Public Works Administration in the 1950s and 1960s. Furthermore, the National Public Works Administration constructed large-scale hydraulic models, the third method, for the same purpose towards the end of the 1940s. And in 1956 for the first time a digital computer was applied. At that moment the civil engineering world applied the four methods for this purpose in parallel.

This case demonstrates that the size of computing activities for the planning of hydraulic works increased considerably in the course of time, and that the people involved developed or applied several new computing technologies. The pressure of local or regional players to get accurate estimates about the consequences of projected hydraulic plans, for instance on the water levels in their regions and on the magnitude of currents, were important for the shipment sector. Moreover the increasing complexity of the hydraulic works, covering large geometrically complex areas, was an important stimulus.[18] As a consequence, an ever greater number of alternatives had to be evaluated with an ever greater precision, so that each party would get insight into the consequences of projected works. The role of rising demand from politicians and lower authorities is emphasized by the fact that many of the engineers involved in the development of new computing technology, who had a background in civil engineering, entered completely new technical fields (such as electrical engineering) to solve this urgent problem.

Advances in the knowledge of computing technologies were of minor influence. Several of these technologies, for instance scale models, were already available before, but were not applied because they were still considered too expensive for the tasks that had to be performed. For instance, Lorentz explicitly rejected the idea of building a scale model to compute the changes in water levels as a consequence of the Zuiderzee works because he considered it too expensive.[19] In other cases, for instance in the case of analogue computing technology, new knowledge was applied readily after its development, but also then the new technology did not provide major price-performance improvements compared to prior technologies. Computer technology provided improved possibilities, for instance to make series of calculations on the same project in a short time, but at a higher price. The new technologies would probably not have been applied much earlier if they had been available, so we can conclude that the application of these technologies depended on the growing demand for computing aids. The limited improvement of computing technology in terms of price-performance stemmed from the fact that computing technology was based on mechanical and electrical knowledge that already existed at the end of the nineteenth and beginning of the twentieth century.

The impacts of the introduction of the new computing technologies were also limited in this period. The computing technology facilitated the evaluation of hydraulic plans and the discussion on these works, but if it had not been available, the evaluations could have continued to be executed manually. Manual computing remained in use well into the 1950s. These machines were not applied outside the field of hydraulic engineering either, so their existence and use did not have wider social impacts. The conclusion is that the main enabler of the growing degree of application of computing technologies in terms of the analysis of this paper were changes in demand, and that the resulting impacts were limited.

The same pattern can be found in the field of government statistics.[20] In the period 1900–60 the Dutch Central Bureau of Statistics (CBS) applied an increasing number of desk calculating and punched-card machines for its operations. At the beginning of the century the bureau performed all computations manually. For large assignments a large number of temporary workers were hired. In 1916 the bureau acquired punched-card machines for the department that compiled trade statistics, which had just been reorganized at that moment. In the 1930s the number of machines started to grow. Around 1950 the bureau counted about 60 machines processing punched cards, which were operated by about 200 employees, processing millions of punched cards each year. By that time employees sometimes characterized the bureau as a 'number factory' or a 'number crunching mill'.[21] In 1960 the bureau introduced the digital computer, an expensive device at that time.

Again demand, in this case demand for statistics, has to be considered as the most important enabler for the growing rate of applications of computing technology. The bureau responded to the growing demand from government agencies and others for statistics. It compiled a growing number of statistics in the course of the century, and the accuracy requirements and the quantity of primary data grew in the course of time.

The fact that around the Second World War government started to pursue an economic policy was an important stimulus for the extension of statistical activities. The application of new technology served this purpose. For instance, compared to punched-card machinery the computer could perform more calculations and could integrate a number of operations in a single programme.

Additionally in this case, new or improved computing technology can hardly explain the growing degree of application of computing technologies. Most technologies would not have been applied before even if they had been available. For instance, the punched-card technology and the digital computer technology from the 1950s most probably would not have been applied by the bureau in 1930, because the smaller workload, the smaller number of tables required, and the lower accuracy requirements of the data-processing operations would not have justified its application, at least not in these quantities. A reason for this is that, as we already mentioned above, price-performance improvements of the punched-card machines were marginal. Punched-card machines became faster and more versatile, but their costs grew faster than labour costs. In 1925 in the Netherlands, a tabulating machine cost *f*2100 per year to rent, approximately equivalent to one-and-a-half years' salary for a machine operator. In 1957 the rent of the machine amounted to *f*20,000 per year, four years' salary.[22] So, price-performance improvements of punched-card machinery cannot explain the growing volume of statistics. The fact that this type of technology remained largely based on traditional electrical and mechanical knowledge was an important cause of the limited improvements in terms of price performance.[23] The same counts for the quick adoption of the digital computer in 1960: it offered extended capabilities, but it was also more expensive and prone to failure. Its application depended mainly on the ever growing quantity of statistical work, not on the new possibilities on offer.

The impact of the application of new technology remained limited, too. Punched-card machinery facilitated the processing of data, in the CBS and elsewhere, but it hardly led to growing levels of computing activities. In the CBS manual computational work remained to be done next to the punched-card operations. And when there were shortages of card punchers in the second half of the 1940s, the bureau started to compile statistics manually that had been compiled with punched-card machinery before. Apparently, the advantages of punched-card operations were not so great. Of course, the machinery had important impacts in the direct labour situation in computing practices, but its wider social impacts were small.

Also, in other applications of early digital computers demand seems to have been dominant. Clearly, growing military demands were important in the genesis of digital computers. However, the advantages offered by digital computers in this period compared to earlier technologies were limited. Whether this is completely true can be doubted, since mathematicians also could apply manual computing methods for such purposes, but computers certainly have facilitated this development. However, here also their wider social impacts were very restricted. In terms of price performance, the digital computer was comparable to prior technologies. It offered improved capabilities, but at a much higher cost than those prior technologies. As a consequence, its adoption has mainly to be explained with reference to growing demands. For instance, the first atomic bombs could not have been developed without computers.[24] But in many applications, traditional technologies could have continued to do the job without much additional costs.

An important reason for the limited advantages of digital computers compared to prior technologies is again the fact that the degree of new technological knowledge embodied in the first digital computers compared to prior computing technologies was gradual, and mainly concerned the composition of different components in the machines that had already been applied in computing technology before.[25] To underscore this point, in 1962 J. Maughly and J.P. Eckert, builders of the famous ENIAC in the US, which is often considered the first electronic computer, commented in an interview in the journal *Datamation* that most of the knowledge that they used in building this machine had already been available ten to fifteen years before.[26]

In the period 1900–60, the dominant factor in the emergence and diffusion of technological paradigms was increasing demand, at least in the fields of technical and scientific computing and data processing. Elsewhere it has been demonstrated that the situation in the field of control was similar.[27] At the same time the societal consequences of the application of the new computing technology were small in this period and, as we will see, minor if we compare them to the period hereafter. New computing technology facilitated the planning and execution of hydraulic works. Computing technology facilitated new forms of control in firms, the implementation of government policy and the development of new technology in fields such as hydraulics. But if the technology had not been available, the computations could have been done by hand, with limited extra cost, as was in fact often done until well in the 1950s for large hydraulic works and even for large statistical projects. So, the technology facilitated changes that, although at a slightly slower pace, would most probably have taken place without the availability of the technology.

ACCELERATING TECHNOLOGICAL DEVELOPMENT

The situation sketched above changed around 1960, when advances in technological knowledge became far more important for the development and diffusion of computing technology. The first digital computers were expensive bulky machines, prone to failure and in most applications offering only limited advantages compared to existing technologies. After about 1960 they started to improve. Until that time the improvement of computing technologies had involved specific features, but at a higher price. Now computer technology improved on many performance characteristics at the same time, such as speed, memory capacity, size, reliability, user friendliness, while the price decreased drastically.[28] This process started around 1960, when manufacturers brought transistorized digital computers on the market.

Regression analysis demonstrates that the price-performance ratio of the computer improved by a factor of about 330 between 1954 and 1984, taking speed and memory capacity as indicators of performance.[29] After that date the process continued. The extent and duration of these price-performance improvements may well be unprecedented in the history of technology since no examples are to be found in the literature that are equal this one.[30]

It is clear that the development of fundamental technological knowledge, particularly in the field of solid-state physics and microelectronics, was important in the development of the digital computer in this period. This path in the development of scientific and technological knowledge was highly rewarding as a resource for improvement of products (chips, computers). Although organizational factors in design and production and improved distribution channels have certainly added to the improved price-performance ratio of the digital computer, the development of technological knowledge was highly important. The possibilities offered by physical nature explain why the improvements were more fundamental and bigger than they are in most other new technologies. Once the solid-state trajectory had started, physicists and engineers saw enormous possibilities for advance. The development of technology knowledge became a strong factor behind improvements of products, and thereby behind adoption decisions of users. In other words, had each new generation of digital computers that was introduced after 1960 been available before, it would in many cases have already been applied. The growth in demand for computing and data-processing technologies, of course, stimulated the diffusion of computers, but if the price-performance ratios of computers had remained at their initial levels, computers would have remained appropriate only for a restricted set of applications.

Compared with previous periods technological developments now resulted in far greater societal impacts. The availability of improved and cheaper computing technology, based on microelectronics, formed an important impetus for organizations to computerize all kinds of data-processing and computing activities, including the development of computer-based management-information systems. The scope and activities of organizations were often changed in turn.[31] The sheer number of computers applied increased far more rapidly than in the previous period. In many applications the digital computer dispelled other computing technologies from the market (although, for instance, scale models continued to exist for quite some time) and new applications emerged. Rapidly developing computing technology became a driving force behind changes in management, and it even made certain management ranks superfluous (Business Process Redesign). Furthermore, in communication, media and transport the computer provoked important changes or became an important facilitator of developments that were already present. People started to apply computers in their homes for text editing, bookkeeping and gaming from the 1970s, creating a large new market for computers and affecting many people's daily life.[32]

In this period the social changes accompanying the introduction of computers, to a large extent, took on the character of 'impacts'. This means that the literature on the information society mentioned above, although expressing a monocausal view of socio-technological development in which technology generates social change, is in fact largely correct as far as its claims are confined to this period and field of technology.

CONNECTIVITY IN THE 1990s

Although it is difficult to give a long-term perspective on very recent history, since the beginning of the 1990s a new period in the development of computing technology seems to have started. Stand-alone computers still improve considerably in terms of speed and capacity, in which advances are largely used to improve user friendliness of software. But the resulting price-performance improvements of stand-alone computers are no longer as fast as before, particularly since the prices of computers are stabilizing.[33] Although knowledge in the field of microelectronics is still advancing, the speed of advance of stand-alone computers is diminishing. It seems that the digital computer as such is reaching its maturity. This would mean that for applications that can be performed by the computer on its own, changes in demand again become more important. The growing attention of computer suppliers to users' requirements since the 1990s confirms this impression.

However, a new phenomenon took place in this period: the convergence of computing and telecommunications technology. In a short period the use of computers for communication purposes, particularly by means of email and the Internet, became popular. Most explanations for this development refer to the availability of Internet technology to a larger public due to the end of the cold war, and increasing overall demands for communication.[34] However, technological knowledge development was an important driver for the convergence of the two types of technologies. The increasing digitization of telecommunications technology and knowledge of the physical connectivity of telecommunication and computer technology were a precondition. Several types of new knowledge were involved, such as knowledge of packet switching, data coding, data compression and cryptology. Part of this knowledge was already developed in the 1960s, for instance for a predecessor of the Internet, ARPANET, but this type of knowledge was different from the period before. Knowledge development in the period 1960–90 concerned fundamental technological knowledge development on product and process technology in the field of integrated circuits. In this period, knowledge development was not primarily fundamental new knowledge, but it concerned applied knowledge of the connectivity of two types of technology. This knowledge, nevertheless, created the conditions for the combination of computers and several types of communication technologies, which offered new possibilities to users of both. On the other hand, scale effects were important in the diffusion of Internet technologies. The local area networks that many organizations established in this period often were not primarily meant for external communication purposes, but to facilitate the implementation of new versions of user software. The packages had to be installed only once on the server, instead of on each individual desktop computer in the organization. These local area networks were an essential precondition for the success of the Internet and other computer-based communication systems. In fact, these local area networks generated scale effects in the maintenance of internal computer infrastructures of organizations and in their communication systems. On the other hand, it is clear that rising demands cannot explain the convergence of computers and communication technologies. If the new technologies would have been offered before, it is most likely that they would have been adopted at that point in time; so the demand for computer communication that became manifest in this period existed already in the decades before. This means that a technological factor, knowledge of the connectivity of computers and telecommunications, and traditional learning effects (scale effects) were important drivers of the new communication applications of computers. And again in this period large social changes were provoked by the new applications of the Internet. New types of information sources, communication patterns, organizational forms of labour, and relations between firms thus originated.

DISCUSSION

The pattern outlined in this article suggests that technology has its most important impacts in periods in which knowledge advances strongly. The advances become evident in improving price-performance ratios of the products based on this technology. The basic idea of this paper is that the advance of knowledge can take place at greatly different speeds, that when the speed is high the impacts of technology are at their strongest, that they then can be considered proper impacts of technology, viz. of technological knowledge development.

Of course, the development of new knowledge and the creation of social change by applying new or improved products remains the result of human action. The conception of the digital computer, the transistor and the microchip are all the result of deliberate actions of the engineers and scientists involved. They could have made different design choices, which might have affected the course of technological development in this field. Moreover, the intellect of the scientists and engineers involved in knowledge and product development, and the level of investments made in specific fields of technology have, of course, affected the speed of development. But these factors alone cannot explain the differences in the speed of the advance of knowledge in different fields and periods. The degree of success of scientists and engineers to advance the state of knowledge depends also in part on the possibilities offered by physical nature. Some roads of inquiry, we may also say some technological paradigms, are more rewarding than others. The microelectronics paradigm was extremely successful, not only through the genius of the people involved but also because nature appeared to offer possibilities for advance. In other fields this may appear far more difficult. An example may be cancer research, a field that only slowly generates improved treatment possibilities. In fact, the role of physical reality in the advance of technological knowledge is the reason to speak of impacts of technology proper, since all other factors are directly related to human action. New fields of knowledge will often offer large possibilities for advance in the beginning, and when they mature the degree of advance decreases. The length of the period of advance is an important factor for the total advance that is realized. In the case of microelectronics this period appeared to be very long, several decades. When the technology matures, only gradual improvement will normally still be made.

CONCLUSION

In this paper I have demonstrated that in the field of computing different periods can be distinguished according to the main enablers of technology development and diffusion and the degree of impacts. I argued that in periods in which knowledge development was the main enabler, technology had its most prominent social impact. The issue of enablers of innovation was treated in a non-categorical sense, making it time-and-technology specific. I demonstrated that three periods formed the scenery of innovation in the field of computing technology: a demand-enabled period between 1900 and 1960, and two knowledge-enabled periods, one between 1960 and 1990, facilitated by knowledge development in the field of microelectronics, and one from 1990 to 2000, and in fact to the present, enabled by knowledge of the convergence of computing technology and telecommunications. In both latter periods the speed of knowledge development was very influential to the improvement in price-performance ratios of IT products, which in turn affected social change. It means that if we consider 'technological revolutions' as periods in which technology exerts a pervasive influence on society, the computer revolution has started around 1960, changed its character around 1990, and continues to the present day. Being still mainly demand-enabled, the invention of the first digital computer did not bring about a new era, but the revolution was brought about by knowledge development that was applied to the improvement of computer technology and the technological knowledge in the connection of computer and telecommunication technology, not primarily by the invention of digital computers. New knowledge was important before the digital-computer technology could result in impacts. This point explains why such a long period elapsed between the invention of the digital computer and its impacts. Such time lags are a more general phenomenon,[35] which may more often share the same result.

In the introduction I mentioned that the issue of this paper is relevant for the history and philosophy of technology. One of the reasons for the very existence of the history of technology as a discipline separate from socio-economic history is the influence of technology on the development of society. The constructivist paradigm has neglected this influence and thus creates the risk that historians and philosophers remain to one side in discussions on the impact of technology on society. While economists recognize the importance of technology in the present period of economic prosperity and are discussing the specifics of this influence, historians and philosophers cannot continue to focus only on the social shaping process of technology, 'deconstructing' deterministic views of technology, while neglecting specific patterns in the evolution of technology which generate major social and economic changes. Dedicating more attention to technological knowledge as a enabler of innovation and the degree to which technology has impacts, enhances the insight into the role of technology in the development of society. The analysis of this paper suggests that the influence of technology on society becomes most clearly evident in knowledge-enabled periods and fields of technology. Determining when such periods and fields occur may be an interesting challenge for the history of technology as a discipline.

Acknowledgements

The author is grateful to Arie Rip, without whom this paper would not have reached its present form. He furthermore thanks Dick van Lente, Wilfred Dolfsma and Onno de Wit for their valuable and encouraging suggestions and comments on earlier versions of this paper.

Notes and References

1. W.E. Bijker *et al.* (eds.), *The Social Construction of Technological Systems* (Cambridge, Massachusetts, 1987). W.E. Bijker and J. Law (eds.), *Shaping Technology, Building Society* (Cambridge, Massachusetts, 1992).

2. M.R. Smith and L. Marx (eds.), *Does Technology Drive History?* (Cambridge, Massachusetts, 1994).

3. Th.J. Misa, 'Retrieving Sociotechnical Change from Technological Determinism', in Smith and Marx, *op. cit.* (2), 115–41.

4. Smith and Marx, *op. cit.* (2).

5. R. Coombs, P. Saviotti and V. Walsh, *Economics and Technological Change* (Houndmills, 1987); D. Mowery and N. Rosenberg, 'The Influence of Market Demand upon Innovation: A Critical Review of Some Recent Empirical Studies', *Research Policy*, 1979, 8: 103–53. The discussion goes back to J. Schmookler, *Invention and Economic Growth* (Cambridge, Massachusetts, 1966). The idea that technology and demand develop in an interactive way is also widespread in neo-evolutionary literature on technology development. See for instance: G. Dosi, 'Technological paradigms and technological trajectories', *Research Policy*, 1982, 11: 147–62; G. Dosi *et al.* (eds.), *Technical Change and Economic Theory* (London, 1988); R.R. Nelson and S.G. Winter, *An Evolutionary Theory of Economic Change* (Cambridge, Massachusetts, 1982); OECD, *Technology and the Economy. The Key Relationships* (OECD, 1992).

6. See, for instance, W. Kaiser, 'What Drives Innovation in Technology?', *History of Technology*, 1999, 21: 107–23.

7. Changes in price-performance ratios of products have been studied in so-called

technological measurement studies. See: A.J. Alexander and B.M. Mitchell, 'Measuring Technological Change of Heterogeneous Products', *Technological Forecasting and Social Change*, 1985, 27(2/3): 161–95; K.E. Knight, 'A Functional and Structural Measurement of Technology', *Technological Forecasting and Social Change*, 1985, 27(2/3): 107–27; P.P. Saviotti, 'An Approach to the Measurement of Technology Based on the Hedonic Price Method and Related Methods', *Technological Forecasting and Social Change*, 1985, 27: 309–34; P.P. Saviotti, P.C. Stubbs, R.W. Coombs and M. Gibbons, 'An Approach to the Construction of Indexes of Technological Change and of Technological Sophistication: The Case of Agricultural Tractors', *Technological Forecasting and Social Change*, 1982, 2: 133–48.

8. H.H. Goldstine, *The Computer from Pascal to Von Neumann* (Princeton, 1972); M.R. Williams, *A History of Computing Technology* (Englewood Cliffs, 1985). A critique of the internalist view has already been given in M.S. Mahoney, 'The History of Computing in the History of Technology', *Annals of the History of Computing*, 1988, 10(2): 113–25.

9. Goldstine, *op. cit.* (8). N. Metropolis, J. Howlett and G.-C. Rota (eds.), *A History of Computing in the Twentieth Century* (New York, 1980). The relatively large attention to military stimuli may be due to the fact that authors such as these have been involved in the military projects that contributed to the origins of the digital computer.

10. M. Campbell-Kelly and W. Aspray, *Computer. A History of the Information Machine* (New York, 1996); P.E. Ceruzzi, *A History of Modern Computing* (Cambridge, Massachusetts, 1998).

11. T. Forester (ed.), *The Microelectronics Revolution* (Oxford, 1980); T. Forester (ed.), *The Information Technology Revolution* (Oxford, 1985); T. Forester (ed.), *Computers in the Human Context. Information Technology, Productivity, and People* (Cambridge, Massachusetts, 1990).

12. J.R. Beniger, *The Control Revolution. Technological and Economic Origins of the Information Society* (Cambridge, Massachusetts, 1986).

13. Campbell-Kelly and Aspray, *op. cit.* (10); P.E. Ceruzzi, 'Crossing the Divide: Architectural Issues and the Emergence of the Stored Program Computer, 1935–1955', *IEEE Annals of the History of Computing*, 1997, 19(1): 5–12; Ceruzzi, *op. cit.* (10).

14. L.S. Reich, *The Making of American Industrial Research. Science and Business at GE and Bell, 1876–1926* (Cambridge, 1985); K.L. Wildes and N.A. Lindgren, *A Century of Electrical Engineering and Computer Science at MIT, 1882–1982* (Cambridge, Massachusetts, 1985).

15. J. Yates, *Control through Communication. The Rise of System in American Management* (Baltimore/London, 1989).

16. S. Bennett, 'The Industrial Instrument – Master of Industry, Servant of Management. Automatic Control in the Process Industries, 1900–1940', *Technology and Culture*, 1991, 32: 69–81.

17. See for a more comprehensive description of this case: J. van den Ende, 'Tidal Calculations in The Netherlands, 1920–1960', *IEEE Annals of the History of Computing*, 1992, 14(3): 23–33; J. van den Ende, *The Turn of the Tide. Computerization in Dutch Society, 1900–1965* (Delft, 1994), ch. 4.

18. In this context the complexity of civil works is a demand factor, since it concerns technological developments in fields other than computing technology, and because broader societal demands were behind this growth.

19. Staatscommissie Zuiderzee, *Verslag van de Staatscommissie Zuiderzee* ('s-Gravenhage, 1926).

20. See for a more comprehensive description of this case: J. van den Ende, 'The Number Factory: Punched-Card Machines at the Dutch Central Bureau of Statistics', *IEEE Annals of the History of Computing*, 1994, 16(3): 15–24; J. van den Ende, *The Turn of the Tide. Computerization in Dutch Society, 1900–1965* (Delft, 1994), ch. 5.

21. J. van den Ende, *The Turn of the Tide. Computerization in Dutch Society, 1900–1965* (Delft, 1994), 162.

22. Van den Ende, *op. cit.* (21), 173–4.

23. After the Second World War punched-card machines became available with vacuum tubes, but these only incrementally affected the price-performance of the machines.

24. A. Nijholt and J. van den Ende, *Geschiedenis van de rekenkunst. Van kerfstok tot computer* (Schoonhoven, 1994).

25. The digital computer can be considered an architectural innovation in the sense of Henderson and Clark (1990): a new combination of existing technological elements. R.M. Henderson and K.B. Clark, 'Architectural Innovation: The Reconfiguration of Existing

Product Technologies and the Failure of Established Firms', *Administrative Science Quarterly*, 1990, 35: 9–30.

26. Nijholt and Van den Ende, *op. cit.* (24), 153.

27. J. van den Ende, 'A History of Real-Time Industrial Process Control in the Dutch Steel-Making Industry', *Real-Time Systems*, 1995, 8(2/3): 215–26.

28. These improvements were much stronger and had different origins than the well-known 'learning effects' involved in the development of technology. See G. Dosi, 'The nature of the innovative process', in G. Dosi *et al.* (eds.), *Technical Change and Economic Theory* (London, 1988), 221–35.

29. R.J. Gordon, 'The Postwar Evolution of Computer Prices', in D.W. Jorgenson and R. Landau (eds.), *Technology and Capital Formation* (Cambridge, Massachusetts, 1989), 77–125. Although the price of first-generation digital computers already started to decrease in the 1950s, I take the first practical application of second-generation computers around 1960 as a starting point for the knowledge-enabled period, since then product knowledge, particularly knowledge in the field of microelectronics, became very important in the improved price-performance ratios.

30. For instance, the literature on technological measurement mentioned before (note 6) gives no comparable examples.

31. N. Venkatraman, 'IT-induced Business Reconfiguration', in M.S. Scott Morton (ed.), *The Corporation of the 1990s* (New York, 1991), 122–57.

32. Y. Aoyama and H. Izushi, 'Hardware gimmick or cultural innovation? Technological, cultural, and social foundations of the Japanese video game industry', *Research Policy*, 2003, 32: 423–44.

33. For stand-alone computers 'performance' relates to the attributes of the pre-1990 period. Of course, in this period computers got additional communication capabilities as a consequence of the convergence with communication technology. In the analysis of price-performance ratios after 1990 these extended attributes have to be taken into account.

34. J. Abbate, *Inventing the Internet* (Cambridge, Massachusetts, 1999), 187.

35. J. van den Ende and R. Kemp, 'Technological transformations in history: how the computer regime grew out of existing computing regimes', *Research Policy*, 1999, 28: 833–51; A. Rip, 'Assessing the impacts of innovation: New developments in Technology Assessment', International Workshop on Social Sciences and Innovation, Tokyo, 29 November–2 December 2000.

Special Issue

The Global History of the Steam Engine

Edited by
Ian Inkster and Patrick O'Brien

This collection is dedicated to the memory and work of Gerry (T.L.) Martin (1930–2004), Inventor, Entrepreneur, Scholar and Patron of the History of Technological Change, and a member of the Editorial Board of *History of Technology*.

As Director of the Renaissance Trust, Gerry Martin sponsored the international Windsor Conferences, a series of three that were convened between 2000 and 2002 by Professor Patrick O'Brien, Centennial Professor of Economic History at the London School of Economics. The first two of these broached the large questions concerning the roles of useful and reliable knowledge in the material evolution of Europe, India and China from 1368 onwards, and examined the variations between states and markets as possible explanations of diverging historical trajectories. In all the discussions Gerry Martin's participation served to focus our wide-ranging arguments back upon artefacts and machines as real and central issues. His refusal to accept Eurocentric assumptions is illustrated in his coining of the term 'useful and reliable knowledge' as a more meaningful category than that of 'science' in exploring the huge complexity of the historical nexus of ideas, knowledge, artefacts and institutions in very different geographical and cultural settings. The present collection represents a selection from the third conference, which took place in the spring of 2002, on the evolution and diffusion of the steam engine in Europe and China.

Introduction – Indisputable Features and Nebulous Contexts: the Steam Engine as a Global Inquisition

IAN INKSTER

With some reference directly to Donald Cardwell, Floris Cohen has written that one 'indisputable feature of Watt's steam engine – the chief technological motor of industrialization – is that the machine is entirely dependent on a prior awareness of the void and of atmospheric pressure, which are insights gained in the course of the Scientific Revolution'.[1] For such a reason, the history of the steam engine must be traced through the giants – Evangelista Torricelli, Otto von Guericke, Blaise Pascal and Robert Boyle amongst others – for its successful evolution depended on a novel creation of 'artificially produced nature' via the air pump – that is the vacuum could only be intentionally produced in a milieu that regarded as both feasible and desirable the creation of what Sorel called 'artificial nature' as a method of investigation beyond that of searching the phenomena of nature as they presented themselves immediately to the senses – perhaps this was the crucial feature of instrument-based experimentation, and perhaps Torricelli and his broken column of water was the initial key, who went beyond Galileo by realizing that it was the pressure of the atmosphere that enabled a suction pump to work.[2]

Now, for stark comparative purposes, we may insert Joseph Needham as of May 1963:

> What constituted the fundamental revolution of the European seventeenth and eighteenth centuries was the inversion of the direction of motion so that force was transmitted not to but from the piston. One may thus justly conclude (if it is not putting too much strain on our adopted terminology) that the great 'physiological' triumphs of 'ex-pistonian' Europe were built upon a foundation of formal or morphological identity laid by 'ad-pistonian' China.[3]

Was there anything in Chinese science or in Chinese culture that might be identified as prohibiting such an inversion?

'To ask why modern science and technology developed in our society and not in China is the same thing as to ask why capitalism did not arise in China, why was there no Renaissance, no Reformation, none of those epoch making phenomena of that great transition period of the fifteenth to the eighteenth centuries.'[4] Here, then, is Joseph Needham's grand global inquisition. Is the answer simply that one (Renaissance or science) caused the other (Reformation or modern technology)? No. For Needham, all this arises because all such phenomena were contained within an even greater bifurcation, that of the Chinese cultural homoeostasis and the European 'schizophrenia of the soul', the grand titration. This is the roundly determining distinction between the two great systems.[5]

In his strikingly ambitious lecture to the Newcomen Society in May 1963 Needham offers a steam-engine version of his general[6] Strong Programme position on technological possibilities in pre-modern China:

> ... it is possible to show that the entire morphology (and some of the physiology) of the reciprocating steam-engine of the early nineteenth century was prefigured in Asian, especially Chinese, machinery, widely used at the beginning of the thirteenth. In order to pass beyond the single-stroke atmospheric stage and the non-rotary rocking beam stage, as Watt, Trevithick and their contemporaries did, currents of design were required which went back centuries earlier than Newcomen. Moreover, these currents were characteristically East Asian.

A great portion of the subsequent lengthy and detailed paper traces such currents on both cognitive and technical levels (if you will, culturally) – these including water-jacket insulation devices, cog-wheels, and steel bearings – but we might here just depict one seemingly crucial instance in more detail, particularly in consideration of the truly stimulating arguments found within this present collection in the substantial papers by Elvin, Cohen and Deng. When considering the question of vacuum we might bear in mind a recent judgement from Kenneth Pomeranz that the:

> Chinese had long understood the basic scientific principle involved [in the steam engine] – the existence of atmospheric pressure – and had long since mastered (as part of their box-bellows) a double acting piston/cylinder system much like Watt's, as well as a system for transforming rotary motion to linear motion that was as good as any known anywhere before the twentieth century.[7]

The Needham perspective, however modified, inspires the background of many a present picture.

Now in the latter case mentioned by Pomeranz – the conversion of rotary to linear motion and vice versa – there seems good reason to view China as easily on a par with, or well ahead of, anything happening in the West. In medieval China the horizontal waterwheel was used for blowing metallurgical bellows. And from an early period these embodied a conversion of rotary to longitudinal motion in heavy-duty machinery, as described by Deng. At this point in his argument Needham is claiming the use of a driving-belt, which may be disputable[8] but does not alter the basic conception. It is accurately and beautifully described in a text on agriculture and rural engineering by Wang Chen in 1313 (with Needham's technical terms!):

> A place beside a rushing torrent is selected, and a vertical shaft is set up in a framework with two horizontal wheels so that the lower one is rotated by the force of the water. The upper one is connected by a driving-belt to a smaller wheel in front of it, which bears an oscillating rod. Then all as one, following the turning of the driving wheel, the connecting-rod attached to the eccentric lug pushes and pulls the rocking roller, the levers to left and right of which assure the transmission of the motion to the piston rod. Thus this is pushed back and forth, operating the furnace-bellows far more quickly than would be possible with man-power.

Needham goes on to emphasize that the system of three parts – the eccentric, connecting-rod and piston-rod – 'has not so far been found in any fourteenth-century European illustration, and occurs only rarely in the fifteenth', and Needham himself seems happy settling back with Leonardo da Vinci in the late years of that century!

But what if we wish to get away from a miscellany of bits and pieces? What if there is an argument that absolutely centres upon one essential feature of the Western steam engine, the material application of the notion of a vacuum? The vacuum-creating single-acting air pump is convention-ally given a Greek root – as is suggested by imagery in the famous Chantrey bust of a classical Watt. What of the double-acting principle – the construction that ensures that the piston does effective work on each stroke back and forth, as embodied in Watt's separate condenser engines in the 1780s? With a full dependence upon the very unusual book by R.P. Hommel, *China at Work*, published in New York in 1937, Needham identifies the thirteenth- to seventeenth-century Chinese box-bellow used in metallurgy as an effective double-acting force and suction pump using the vacuum as its central principle. Writing in 1842, Thomas Ewbank identified this Chinese technology with its two cylinders ingeniously combined into one as effectively a Boylean air pump, and of a construction 'identical with that of the steam-engine; for let it be furnished with a crank and flywheel to regulate the movements of its piston, and with apparatus to open and close its valves, then admit steam through its nozzle, and it becomes the double-acting engine'.[9]

If this still seems a little way from the double-acting steam-powered engine, then Needham (following Tredgold) suggests an evolutionary passage: in late eighteenth-century Britain 'the double-acting principle was embodied first in "piston bellows" worked by steam prime movers, and then in the prime movers themselves, the one use leading to the other by a natural transition.'[10] Thus for Needham China had the working essentials at a very early stage, and he strengthens his argument with an independent explanation of the early Chinese transition from bellows to pistons in his belief that pistons arose through a gradual reduction of the leather component of the bellows:

> Perhaps the leather skin of the pot or drum bellows was induced to turn into a piston at a surprisingly early stage because of the presence in one strand of Chinese culture of a simple piston and cylinder which was not designed to push either air or water from place to place, but to heat air to an ignition point by a sort of adiabatic compression, and so to make fire.[11]

Such applications were truly ancient, originating possibly in Malaysia.

This seems to need consideration if we wish to compare or contrast the cognitive and technical resources for the steam engine. One large point of doubt – does the box bellow involve not so much a conception or use of a vacuus *space* (or column) as the use of vacuum or suction power? Thus it seems possible that these Chinese technologies did not demand realization of the possibility of a vacuum that actually takes up space nor, thus, the formulation of atmospheric pressure. The example seems almost exactly analogous to the operation of the pre-Christian force pump so vividly described by Hollister-Short in his 1993 paper in *History of Technology*.[12]

Several presentations and responses at the workshop rejected the historical veracity of any such direct artefactual contrasts. The steam engine was no collection of components but a physical and cultural outcome of a host of features specific in their combination or configuration to northwestern Europe. As Nathan Sivin emphasized in his comments on Elvin's paper, taking templates designed for evaluating the European scientific tradition and using them to decide what passes muster in China has been at best a distraction in the past, and we would perhaps do better to devote our efforts to the patterns actually inherent in Chinese technical activity. This was a consideration in Sivin's own paper on the railways, as well as a point developed by other participants.

In any case it could be argued, as we have already hinted, that creative moments and realizations are one thing, the ability to react to new circumstances another. If the occasion of crisis is the window to opportunity, then consideration of the steam engine in comparative or global history requires a move beyond its origins and towards the resources required for positive and sustained reaction to advancements elsewhere and for diffusion within follower systems, in the understanding that the term 'resources' might embrace an array of intellectual, cognitive and cultural assets. If we were to accept all of this then we would be encouraged to conclude that technology *reflects* rather than causes the major changes in the advancement of the material world. By this argument, the steam engine or any other progressive and heroic breakthrough is rendered irrelevant as a major explanatory variable.

Notes and Footnotes

1. Floris H. Cohen, *The Scientific Revolution. A Historiographical Inquiry* (Chicago and London, 1994), 527; see D. Cardwell, *Turning Points in Western Technology. A Study of Technology, Science and History* (New York, 1972).

2. The reference here to Georges Sorel is to his *Les préoccupations métaphysiques des physiciens modernes*, Cahiers de la Quinzaine, série VIII, no. 16 (Paris, 1905).

3. Joseph Needham, 'The Pre-Natal History of the Steam Engine', *Newcomen Society Transactions*, XXXV (1962–3), 49.

4. *The Grand Titration* (London, 1969), 176.

5. For some discussion of which, see Cohen *op. cit.* (1), 461–82, and compare this with Elvin essay in the present collection.

6. *Newcomen*, 3–55, quote p. 3. For good summary positions of a truly massive and yet continuing corpus see Joseph Needham, *The Grand Titration*, 14–54; *idem.*, 'Science, Technology, Progress and the Break-through: China as a Case Study in Human History', in Tord Ganelius (ed.), *Progress in Science and its Social Conditions* (Oxford and New York, 1986), 5–22.

7. Kenneth Pomeranz, *The Great Divergence. China, Europe and the Making of the Modern World Economy* (Princeton, 2000), quote pp. 61–2.

8. See Needham, *Grand Titration*, 32 and A. Burstall, *A History of Mechanical Engineering* (London, 1963). The latter is also an excellent source for an even earlier interconversion by means of cogged wheels operating on a slot-rod.

9. T. Ewbank, *A Descriptive and Historical Account of Hydraulic and other Machines for Raising water, Ancient and Modern* (New York, 1842), 271.

10. Thomas Tredgold, *The Steam Engine* (London, 1827), see plate 13; Needham, *Newcomen*, 23.

11. *Ibid.*, Needham, 24.

12. Graham Hollister-Short, 'On the Origins of the Suction Lift Pump', *History of Technology*, 15 (1993), 57–75. As the author makes clear, the true suction pump of the 1400s, with its valved piston, also utilized atmospheric pressure but it is of note that he also astutely judges that a formal conception of atmospheric pressure was not needed of the experimenting craftsmen, e.g. p. 59.

Some Reflections on the Use of 'Styles of Scientific Thinking' to Disaggregate and Sharpen Comparisons Between China and Europe from Sòng to Mid-Qing Times (960–1850 CE)

MARK ELVIN

INTRODUCTION

In his *Styles of Scientific Thinking in the European Tradition*,[1] published in 1994, Alistair Crombie suggested that there were six styles of thinking in the West that had originally been distinct but eventually combined with each other, as they developed, to form the conceptual approach that characterizes 'modern' science. The underlying argument was that expressed by Alessandro Piccolomini in the sixteenth century, that 'any discipline should receive its name rather from its way of demonstrating than from its subject-matter'.[2] In other words, according to what was felt to be persuasive, and why.

The six styles were as follows. The 'postulational' was that based on axioms and definitions, and articulated through the deduction of theorems by means of formal rules of derivation. Its best known example is Euclid's geometry. The second was the 'experimental'. It was the exploration of immediate causes by the control of a small-scale physical situation where all inputs but one are held constant and this other input varied, and the resulting differences in output observed. 'Thought-experiments' are also possible, and can be evidence for the presence of this style of thinking. The third was 'modelling'. It involved the creation of a model, either actual or imaginary, that reproduced the behaviour of some part of nature in certain respects, such as the rotation of the celestial sphere. In the physical version, it was normally of a smaller size than the reality, with attendant scale-related problems. The fourth style was 'taxonomic', that is to say the structured classification of the members of some coherent part of nature, such as plants or animals, in accordance with a set of identifying characteristics. Linnaeus may be thought of here as defining a canonical example. The fifth style was the analysis of causally opaque events that were similar in some respects but different in others, in terms either of the calculus of probabilities or of other statistical concepts. In other words, 'probabilistic' thinking. Examples are the throws of a die, or ages at death. The sixth style was explanation through 'historical derivation'. Its earliest developed Western example was the reconstruction of the ancestry of the family of Indo-European languages, as regards the sequential transformations of semantics, grammars and pronunciations.

All of these styles existed in Europe in elementary form prior to the seventeenth-century 'scientific revolution', either in classical antiquity or the European middle ages, with one exception. Probabilistic thinking was an early modern creation. It was foreshadowed in the sixteenth century by Cardano, and had its formal birth in the middle of the seventeenth with Pascal and Fermat. The 'revolution', in the perspective of styles of thinking, thus appears as the outcome of an increase in the pace and intensity of the development of these styles of thought, their combination, the widening of the communication of results, and an ever more integrated coordination of research agendas. Thus underlying the 'classical' Darwinian conception of the evolution of species was a new combination of the 'taxonomic' and the 'historical' styles. That a historical explanation requires additional considerations, in this case the expansion of the British Empire (whose navy made the *Beagle*'s voyage possible) and generations of prior work by imperial scientific officers on such topics as the flora of previously unfamiliar places like St. Helena,[3] plus the stimulus of Malthus's ideas, goes without saying. The styles are only the abstract structures that define what is accepted as persuasive.

One can be critical of this approach as a *general* method of understanding the birth of 'modern' science, but it can have real heuristic value, and Rob Iliffe has written an important review-article of Crombie's work in this vein.[4] My objective here is not to defend Crombie's vision as a whole. It is to suggest that it can be useful as a way of approaching the comparison of the scientific traditions of China and Europe up to the period when 'modern' science crystallized. By using the disaggregation provided by Crombie's styles of thinking, and taking each of them separately, it is possible to look more precisely at what was, and what was not, shared between China and Europe; and to avoid having to use the term 'science' as an analytical category, which can mire the exercise in conflicts between *a priori* commitments to particular viewpoints.

I shall show that all Crombie's styles of thinking, except the probabilistic, can be found in premodern China, though usually in a rudimentary form. Some off-the-record rule-of-thumb probabilistic thinking fairly certainly also existed, since at least Song times, as a guide to professionals in the aleatoric arts, though no texts have been located that prove this other than inferentially.[5] The challenges which this disaggregated approach present frequently emphasize even further the elusive nature of what is already recognized as a complex conundrum.

If we consider the period from about 1600 to 1800, what we find in China is on the whole only the most minimal *improvement* in any of the styles of thinking. Thus, in taxonomy, no advance in systematization is made after Lî Shízhen's *Bêncâo gang mù* [Pharmacopoeia arranged by headings and subheadings] of 1596.[6] The standstill was nonetheless not total. In the style of 'historical derivation' the work of the Qing dynasty scholars on reconstructing ancient Chinese pronunciation showed savants capable of formulating a 'program' of work in the Lakatosian sense, and carrying it out with good intercommunication between participants.[7] Thus purported explanations must distinguish in the future between different outcomes for differing styles.

If, in Spenglerian mode, we lay stress on differences in cultural conceptions we have to confront the implications of the presence of close-to-identical ideas about truth in both cultures in *some* domains. An example is the causes of eclipses, which provoked a crisis in late-imperial Chinese meteorology as improved predictability made traditional *moral* causality implausible.[8] Another is the unavoidable identity of the patterns of frequencies of outcomes when tossing dice or coins. In the latter case we even have an overlap in the technology, since China borrowed the six-sided die from West Asia, developed it in Northern Sòng times into dominoes, and afterwards re-exported the latter to the West, which added the blank face. The use of sale by gambling, notably in the Southern Sòng, required that shopkeepers – in order to survive – had to have at least a rough knowledge of odds. Perceived validity and invalidity in this domain thus had to be, or become (for social-darwinian reasons), the same for anyone anywhere in the world where it existed. Gambling with dice and coins, common in premodern China, has the additional feature that it is close to being a constantly repeated experimental test of theory. Casino-style gambling (with a 'bank', stakes and payoffs) was institutionalized in China by Southern Sòng times or earlier; awareness of the frequencies would have been essential to the professionals who ran it. At the same time there is no sign that experience in general disproved 'unscientific' notions among the client gamblers, who continued to believe in divination, personal luck and the favour of spiritual beings.[9]

Focusing on a given style immediately throws up points like these. The last section but one also discusses what might be seen as yet another style of thinking, the 'mathematization of nature'. I prefer to see this as a *combination* of the experimental and the postulational styles, and it is interesting that Zhu Zàiyù, the sixteenth-century expert on musical acoustics whose work is discussed there, saw it in this way. He commented on his theory of equal-temperament tuning that, to grasp it, scholars would have to be well acquainted both with acoustics and with mathematical calculation.[10]

Historians of China have long known that simple economic arguments crash when applied to differences in scientific advances between China and

Europe in this period. China was the leading economy in the medieval world; in the period from 1600 to 1800 it still had one of the most advanced economies anywhere: highly commercialized and monetized, entrepreneurial, and competitive to a degree that startled the Jesuits, and with yield-to-seed ratios in some areas that dwarfed those in Western Europe at this time.[11] The social penetration of printing was probably not greatly different from that in contemporary Europe,[12] and the art was more than half a millennium older in China. This in itself obliges recent printing-orientated arguments to undergo redefinition.[13] However, when we look through the various styles of thinking in China during these two hundred years, not one of them seems to have benefited from these various economic factors, or even to have been strongly linked to them. This prompts the converse suspicion that in the West – given an adequate basis of productivity, literacy and numeracy, and the physical means of communication – the early-modern advance of the styles was in no case crucially linked in a *general* way with the rising level of Western economic and even technical prowess. The ancient or medieval origins of all but one of the styles, and their early growth, are consonant with this, but it is still a disconcerting hypothesis to have to consider. Specifics can be a different matter.

In counterpoint to this we have also to consider whether 'science' and 'technology' in this age were more like branches growing off the same cultural trunk than separate trees. To give an example, when De Bélidor, the French hydraulic engineer working in the first half of the eighteenth century, observed the differing efficiencies of various shapes of *moulins à chapelet* – so-called 'rosary-bead pumps', derived from the Chinese chain-linked square-pallet trough pumps – he used an abstract geometrical analysis to determine the optimum proportions for pumps with troughs canted at specific angles. He came up with a quantified theoretical version of the empirical Chinese solution to the same problem: to maximize the work done for a given input of energy, use pallets with a lower ratio of height to breadth for flatter angles of lift.[14] So was De Bélidor working in the domain of science or technology? We would tend to say, in his case, both, but the Chinese only in the second. But surely what was critical was a difference in the styles of intellectual culture, which affected both.

Then again, to what domain belonged the fantastic Western machines shown in the sixteenth- and seventeenth-century 'Theatres of Machines and Instruments' concocted by writers like Ramelli?[15] A typical such device was a waterwheel turned by water flowing out of an upper reservoir into a lower pond, and, as it rotated, turning an Archimedean screw that lifted the water back up from the pond to the reservoir again, from which it endlessly flowed down *ad infinitum*. Were these graphics, probably inspired by the works of Archimedes, recovered in the middle of the sixteenth century, a form of art, or of *Gedanken* technology (that is, the exploration of possibilities in the mind, rather than in actuality), or of science? Probably all three. Besides purporting to show perpetual-motion technology they were art in the same sense as Martin Escher's drawings, which also appeal to aesthetic sensibility. But perhaps the most stimulating point about such a picture was the challenge to discover why a real-life version not only did not work, but *could not*; and this was science. Chinese technical drawings could be elegant and clear, or garbled by imperfect transmission and imperfect understanding, but never inspired by this sort of fruitful science-fiction.

So, having argued earlier for at least some unavoidable general truths irrespective of culture, such as outcomes when tossing dice, we are back with culture again. The early-modern Europeans, but *not* the Chinese, were acquiring a sort of X-ray vision that could see, not the ghost in machines, but the geometry. The second diagram in the Axioms section of the *Principia* is in a way an example of this.[16]

It is necessary, however, to be wary about another seeming cultural 'solution'. This is that early-modern Western culture was unique in its capacity to invent mensurable but imagined entities such as 'force', 'momentum' and, slightly later, the generalized concept of 'energy' with its interchangeable incarnations, an idea whose weirdness now fails to surprise us only because of its long familiarity. (It will be recalled that Newton cast his physics mostly in terms of the *ratios* pioneered in classical antiquity because he was uncomfortable with the coexistence of differing 'dimensions' in a single formula.) The West was supreme in this domain, but, as often, there was a weak Chinese parallel. The distinction between 'position-power' (*shì*) and 'strength' (*lì*) had been familiar since at least Hàn times. One early and crude definition, given by Wáng Chong, illustrated it by the difference between the greater strength of an ox relative to a mosquito, but the greater 'position-power' of the mosquito because it could move and sting the beast at will. This term later became commonly used in Chinese hydraulic literature to describe a property of flowing water that it had, not just because of its celerity, but also because of its location.[17] In the hands of Sòng Yìngxing in the seventeenth century it appears as a proto-idea of energy, without potential and kinetic being differentiated:

> When a waterfall in the high mountains falling over a vertical cliff of a hundred fathoms in height strikes the shallow water in a deep gorge below, those who hear it are so affrighted at the sound that they [feel as if they had] lost their souls. Yet, if one leans a broken jug [full of water] on its side, or fills a ditch with water to its brim, one does not perceive any [such] sound. The water is identical, the pouring is identical, but the sound produced is different. Why is this?
>
> I would say that this is [a case of] what one calls 'the position-power of the matter-energy' [*qìshì*]. The matter-energy obtains position-power [from its height] and so sound is forthcoming from this. If it does not have its position-power, the matter-energy will be very weak.[18]

Again, was the intensifying taste in the West, since later medieval times, for innovation in styles of art, architecture, painting, music, and fashion in clothing, of use in loosening fixed habits of thought in a way that was beneficial for the development of new scientific ideas? Especially, perhaps, when the rate of detectable change dropped to under a human lifetime? If so, was it distinctive? The answer is unclear, but here too, there were weak parallels in China. These included crazes (such as a passion for the cultivation of out-of-season flowers in heated greenhouses) and, in the lower Yángzi region, publicly visible fashions in women's dress, starting in the second half of the seventeenth century.[19]

The pattern in many domains of what might be called a 'weak but seemingly unfruitful resemblance' in China to the early modern West time and again renders the positing of sharp contrasts untenable. This leads to tediously cautious phrasing, but has the advantage of ruling out over-easy explanations in terms of one or another 'magic factor'. An important instance is the question of the 'university'. In China a state university existed in some form since the second century BCE (with the transient establishment of provincial colleges in the fifth century). It flourished most notably under the Táng and, above all, the Sòng, when it was known as the *Tàixué*,[20] with examinations in mathematics and medicine. The argument of Grant and others that 'universities' were the key medieval factor in the early growth of Western science will not stand up in *simple* form to the Chinese comparison, though I suspect that if it were reformulated in terms of the differing natures of the respective institutions of higher education it would still retain pertinence.[21] A distinctive feature of the Míng and most of the Qing dynasty, and, crucially, of the period from 1600 to 1800, was that there was no longer such a substantive institution. There were however numerous 'academies' (*shuyuàn*), though they were predominantly 'crammers' for the imperial examinations for entry into the bureaucracy.[22] In what sense was an 'academy' not a 'university'? This is a characteristic case of where an argument with some weight needs rethinking in the light of the Chinese experience.

If it is accepted, in the light of the evidence presented below, that premodern China had all the Crombiean styles of scientific thinking in some explicit form, with the exception of the probabilistic, it follows that what we have to explain, in a positive sense for early-modern Europe and a negative one for China under the late Míng and the Qing, is less an origin than the presence, or an absence, of an *accelerating* process of development of pre-existing styles of thought. Even as we say this, we have to recall that Bachelard long ago showed, in his *La formation de l'esprit scientifique*, that until the close of the eighteenth century there was still an only gradually diminishing flow of whimsical, bogus and frivolous science in Europe. This was perhaps typified by the efforts at the Académie Royale des Sciences to sublimate elephants' ears, just to see what happened.[23]

Yet something decisive had occurred. This was symbolized by the moment when Della Porta, the exponent of 'natural magic' who was for long the dominant figure in the Accademia dei Lincei,[24] and who *can* be paralleled by Chinese scholars of the same time like Xiè Zhàozhè, author of the *Wûzázû* [The fivefold miscellany],[25] was replaced in this role by Galileo who can *not*.[26]

From the point of view of this essay, the key difference was the late-imperial Chinese inability to make any but one or two of their own scientific seeds continue to grow in their own right. This had consequences. As Kuznets pointed out, the *sustainability* of modern economic growth depends on modern science.[27] Otherwise it will be just a passing 'efflorescence', as Goldstone has characterized the economic upsurge in Sòng China that failed to continue its flow of inventions and innovations in mechanized and mass manufacture.[28] The issue is thus also of central importance to understanding the genesis of our own world.

THE POSTULATIONAL STYLE

The postulational style needs three main elements: (1) concern with the terms used *as* terms, especially their definitions; (2) the creation of an articulated structure on the basis of these terms; and (3) the criterion of coherence, or the avoidance of the derivation of contradictory conclusions by different routes through the same system. (Hence the shock effect of Russell's paradox on modern mathematical logic, requiring what intuitively seem artificial barriers on certain constructions.[29]) The best-known Chinese attempt to formulate such a system was the creation by the Mòhist logicians in pre-imperial antiquity of what we would now see as a form of Boolean logic.

Take as an example the precision of their use of markers or demonstrative pronouns. In modern terms it amounted to saying that if, in a valid statement, you substitute x for y, then *every* occurrence of y must be so changed, and *all* the x's (if any) in the original statement changed to something else distinctive. Otherwise the validity of the transformed statement cannot be guaranteed.[30] The Mòhists also gave testable definitions. Thus 'circular' meant 'having the same lengths from a single centre'.[31]

They grappled with the concepts of membership of a set, the structure of a set, and of identity. The passage that follows shows their attempt to clarify the condition determining the identity of two composite sets (which is most simply expressed as that a member of either is always a member of the other). The argument is overtranslated to make it as intelligible as possible. Chevron brackets '<...>' indicate the elements of which the sets are composed, and square brackets '[...]' enclose my commentary. The translation should also initially be read in a strictly formal way, so that when the word 'is' occurs it is not automatically interpreted in the modern sense of '='. At the end I use quasi-modern notation to determine the point at which the logic breaks down. The numbering of sections has been added; likewise the italics and underline for emphasis and the bold type to highlight the components of the Mòhist analogue of a propositional calculus.

> 1. The grounds for regarding as **false** the statement that 'the set of < oxen **and** horses > *is not* [identical to] the set of < oxen >' are the *same* for those as regarding it as **true**. The explanation for this paradox lies in the way composite sets are aggregated.

2. **If** some of the members of the composite set [<oxen **and** horses >] **are not** oxen, **and if** the statement that 'it **is** [i.e., has some members in common with] < **not** oxen > is therefore regarded as **true, then**, – though some of its members **are** of course **not** oxen – since some of them **are** indeed oxen, the statement that 'it **is** [i.e., has some members in common with] the set of < oxen > may also, *on the same grounds*, be regarded as **true**.

3. **If**, for this reason, we regard as **not** being **true** *either* the statement that 'the set of < oxen **and** horses > **is not** [identical to] the set of < oxen >', *or* the statement that 'the set of < oxen **and** horses > **is** [identical to] the set of < oxen >', **then** – since a proposition *must* **either** be the case **or not** the case – we must *also* say that the statement that 'the statement that "the set of < oxen **and** horses > **is not** [identical to] the set of < oxen > "' is **not** to be regarded as **true**' is *also* itself **not** a permissible proposition.

4. *However*, **if** the set of < oxen > is **not** a two-fold composite set, **and** the set of < horses > is **not** a two-fold composite set, *but* [= 'and'] the set of < oxen and horses > *is* a two-fold composite set, **then** there is no difficulty in accepting as **true** the statement that 'the set of < oxen > **is not** [identical with] that of < **not** oxen >', and the set of < horses > **is not** [identical with] the set of < **not** horses >', *yet* the set of < oxen **and** horses > **is not** [identical with] the set of < oxen > **and** [also] **not** [identical with] the set of < horses >.[32]

This rendering requires taking the words *fei niú*, literally '[is] not ox', in the third sentence as 'has some members in common with < **not** oxen >', but per contra as 'not identical with ' in the final sentence. The exposition in the original is thus faulty.

The argument can be made formally rigorous in a bizarre but revealing way. Take '*x*' to mean 'a separable double entity consisting of one ox and one non-ox', and '**e**' to mean 'provides *in any way we wish* a single unit element for'. Give '**E***x*' the usual sense of 'there exists at least one element *x* such that', and define '$\approx$' (which we may term 'Mòhist identity') between two sets A and B by

If (**E***x*) ((*x* **e** *A*) **and** (*x* **e** *B*)) **then** $A \approx B$.

If A and B are, respectively, < oxen > and < not oxen >, this interpretation of the symbols leads to a contradiction. Depending on how we select a single unit from the double-unit entity *x* (and we have given ourselves the right to do it either way we wish, 'ox' or 'not ox'), we can derive either '$A \approx B$' or '**not** $(A \approx B)$'.

The point of the original argument is clear: Mòhist identity can only hold without contradiction between sets containing identical types of elements, like the set of < oxen >, but not between sets that consists of more than one type, such as < oxen **and** horses >. This is not a definition of identity that we would currently accept as useful, but the drive towards rigour is impressive. The passage also shows the implausibility of the views of certain sinologists and others who see the Chinese *language* as somehow incapable of expressing or handling abstract concepts.

After this impressive debut, the Chinese style of postulational thought soon became thin to nonexistent. The Buddhist logic (*yinmíng*) that came in from India during the sixth and seventh centuries CE, had a well-defined structure and a strong concern with proof, or 'that which can establish' (*néng lì*),[33] but its influence, though not negligible, seems not to have been widespread. We need to bear in mind, though, that as Chemla has shown, the systematically built algorithmic structures of Chinese mathematics created early in the first millennium CE, had many of the properties of the postulational style.[34]

The postulational approach regained some strength in the Qing period, following the introduction of elements of European mathematics by the Jesuits. Mínggantú, who wrote a book in 1774 dealing with infinite series and the calculation of the value of π, giving demonstrations of his results, spoke of how initially, 'unfortunately I had only the formulae, without their meaning, and I feared that when people saw the result but not *the tool that allowed one to reach it*, they might have doubts.'[35] Engelfriet, discussing the introduction into China by the Jesuits in the previous century of the first six books of Euclid, has also stressed how restricted any detailed knowledge of Euclid was *in Europe* until the later part of the sixteenth century, and how his postulational methods had tended prior to this to be associated more with philosophy rather than science.[36]

After the fading of Mòhism in Hàn times, Chinese culture did not generate any sustained strong interest in pure postulational thought, but retained a capacity to handle it if the motivation arose.

THE EXPERIMENTAL STYLE OF THINKING

Experiment is most simply defined by looking first at what it is not. It is different from ordinary observation, which can be strongly influenced by cultural preconceptions. Thus Xiè Zhàozhè, on a voyage to the Liúqiú Islands, *observed* dragons:

> Thunder, lightning, rain, and hailstones all fell upon us at the same time. There were three dragons suspended upside-down to the fore and the aft of the ship. Their whiskers were interwound with the waters of the sea, and penetrated the clouds. All the horns on their heads were visible, but below their waists nothing could be seen.[37]

This was a collective sighting, and no doubt those who thought they were seeing dragons influenced each others' perceptions. On other occasions Xiè could be an acute, accurate and sceptical observer. Thus he also wrote:

> The saying has been handed down that after the [winter] solstice snowflakes are five-pointed. But every year as the winter has moved into spring, I have gathered snowflakes and looked at them. All are six-pointed. Not one or two out of ten is five-pointed. Thus one can learn that the old sayings are not all of them entirely valid.[38]

In this case, there is the added element of a high frequency of repetition.

The distinction between observation and experiment may seem self-evident but even historians of science forget it at times. Thus Needham wrongly describes the account by the sixteenth-century pharmacist Lî Shízhen of how he cut open the stomach of a pangolin to see if it ate ants as describing 'the results of experiment'.[39] Lî's assertion that repeated clinical observations showed that the soya-bean would not help cases of indigestion or poisoning unless taken together with 'liquorice' [*gancâo* = *Glycyrrhiza glabra*] comes closer, but the controlled manipulation of the situation that defines 'experiment' is still missing.[40]

More tightly constrained than observation was 'verification' (*yàn*). This typically applied to forecasts in dreams and meteorological and astrological portents. Xiè wrote:

> When the people of the present day read of the dreams recorded in the historical annals, most of the latter were 'verified' [as we say colloquially, 'came true'] ... Now people dream every day. If these [few that were recorded] were the only ones to be verified, the dreams that did *not* come true must have been innumerable.[41]

Here he is not just observing but also asking the same implicit question of each item in a body of evidence: are dreams predictive or not predictive? His answer is a statistical 'no'.

The emperor Yongzhèng in the second quarter of the eighteenth century was a believer in meteorological justice:

> If good people are numerous, and bad people few in number, then Heaven will send down good fortune upon them. ... If bad people are in the majority and good people in the minority, then Heaven will send down punishments upon them. Even those who are good will be affected by its calamities.[42]

This principle had local application, since 'We see, in departments and counties not far distant from each other, that the quantities of moisture provided by the rainfall are different, and that some have good harvests and others dearth.'[43] It could also be that 'the level of the harvest will correspond to the [moral] level of the governor-general and governor.'[44] In 1729 he issued this decree to the grand Secretariat:

> We have already determined that the differences between the various localities in seasonality, and in rainfall and sunshine, are due either to defects in the government at Court, or to local officials failing in their duty, or else to the customs of the people being corrupt, and their heart-minds false and ungenerous. ... There were abundant harvests everywhere last year throughout the province of Zhílì [modern Hébêi]. Only three administrative subdivisions, Xuanhuà, Huáilái, and Bâoan, missed being fertilized by the rains. We thereupon entertained suspicions toward the officials and commoners of these localities, fearing there might be causes for them to have called this down upon themselves.[45]

In the autumn he received a report from an official that there had been a long-standing quarrel between the bannermen (hereditary soldiers) and civilians in this area over the use of a water channel. The official had told them to pay attention to the emperor's concern, after which they resolved their differences, and there was copious 'auspicious snow'. The emperor regarded this as having empirically validated his views. Here a sort of moral epidemiology, focused on examining systematically what covaried with what, developed an element of proto-experiment represented by the official's intervention in reshaping the situation, which produced a new outcome. Underlying this was the testing of an implicit hypothesis: unseasonable weather was caused by immoral behaviour. There were aspects of political semi-charade in the emperor's position, but the logic of his thinking is significant.[46] Yongzhèng's quasi-programme was abandoned under his successor, Qianlóng, presumably because it was not producing believable results. But it would have been ideologically and politically unthinkable to have proclaimed this publicly.

Our foray into the history of moral meteorology reminds us that, while fruitful experiment requires asking a well-formulated question, this question has either to be soluble in more than a negative way, or lead to one such.[47]

What was lacking in this last case was effective repeatable control of the causal inputs and the outputs. Here, conversely, we should note that every skilled cook, gardener or craftsman in China was in one dimension – but usually one only – an experimenter who strove to achieve exactly this. Xiè Zhàozhè recorded a number of cases that illustrate this point. Of the famous poison *gŭ*, made from insects, he said that it needed to be 'tested on a person' once it had been prepared:

> If there is no stranger passing by, then assure yourself of the [poison's suitability] on some member of your family. The person who has ingested this poison will experience a twisting pain in his guts and vomit. All his fingers will go black. When he chews beans they will have no savour; and if he puts alum in his mouth it will not taste bitter. This is the way [the poison] is experientially verified.[48]

Mercifully, there was an antidote.

Horticulture was also a nursery of proto-experimental control:

> These days flowers out of season are among the items regularly presented at Court. They are, moreover, all confined to [semi-subterranean] cellars in the earth, where they are surrounded on all sides by fires so as to force their growth. Thus there are peonies even in the depths of winter ... In fact, such items out of their proper season are in conflict with the regularities of Heaven and Earth.[49]

Such practices were halfway to the Baconian 'torture' of nature to make her yield her secrets, but only halfway because no question was being asked.

A more heavyweight example is the deduction from the production of mercury from cinnabar (HgS) and the *apparent* reverse production of cinnabar from mercury that these materials embodied the alchemical secret of changelessness within change. Cinnabar itself was used to give scarlet lacquer its colour, and as an ingredient in various cosmetics. One of the uses of mercury was in widely sold pills thought to banish nightmares, and it was a traditional component of elixirs designed to confer longevity or immortality. Hence it was the foundation of a substantial industry. In his *Book of Gùizhou*, Tián Wén described in detail the methods of distillation or sublimation in early Qing times.[50] Although these were practical technology they can also be seen as implicitly part of an apparatus for experimental research:

> There are large stoves and small ones. Cauldrons likewise come in two sizes. Large ones can contain just over twenty litres of ore.[51] They are separated into ten layers, which are loaded in sequence with a layer of chaff and bran between each of them. Ashes from which mercury has been recovered are spread on top of these with a ladle so that they form a shallow depression, low in the middle and higher around the circumference. The stove is covered with an inverted cauldron, along whose rim is smeared clay into which salt has been kneaded [in order to seal the join]. Once this assemblage has been built, it is fired. It usually takes a day and a night for the mercury to form droplets, and the droplets to become suspended pearl-sized globules, trembling and shimmering, all of them having risen into the belly of the inverted cauldron serving as a cover.

Cinnabar converts to mercuric oxide which then decomposes at about 500° C, yielding mercury. At that heat the salt in the clay may also have formed a glaze that improved the impermeability of the seal, so stopping the gas from escaping. Packing in separated layers may have helped the gaseous mercury to rise. Once the gas had passed through the ashes it would have condensed on the underside of the top of the cauldron and then dropped back into the hollow depression that thus served as a trap. This process has similarities to some of the German methods described in the *De Re Metallica* of 1555.[52] The text giving the second procedure is not so easy to understand:

> In the case of the small stove, it is filled with small lumps of the ore to be heated and stones, interlayered with each other, and then covered with a sieve-like cover made of bamboo splints, and smeared with bean paste. There are four holes in this to draw off the vapours. These holes connect in closed, circular fashion with the apertures of distilling pipes. The top is covered with small fired tiles the gaps between which are also made impermeable with salted clay. It is then fired and the operation should be finished in the time it takes to burn a stick of incense. The mercury rises up toward the fired tiles, where it condenses and flows out through the holes into the pipes. After it has had time to settle, it is decanted into pigs' bladders and bound, after which it can be carried long distances. ... Those who open either the cauldrons or the fired tiles always hold a leek in their

mouths, or juice from meat pickled with its bones. If they do not do this, their teeth will drop out when they encounter the vaporous energy-matter.

Tián ends with the alchemical theme, the changelessness within change that was the embodiment of the secret of immortality within mortality: 'If mercury that has already been made in finished form is distilled, it can once again be made into cinnabar [when it cools]. *It does not forget that which it fundamentally is.*' But, as is well known, this reversal was an illusion. The reverse process, needing from between 300° to 350° C, would have made red mercuric oxide, using oxygen from the atmosphere, and not cinnabar crystals, as the sulphur needed would largely have been lost after the initial processing. This oxide could have then be turned back into mercury at 500°, apparently demonstrating changelessness within change. At the primitive level of chemical understanding described here the oxygen drawn from the air was invisible to the mind's eye.[53] Though no one, seemingly, asked and then pursued the question of the later incarnations of the material, 'Are they *really* cinnabar?', we are close to authentic experiment.

The formal criteria for the experimental style of thinking can be met by a trivial example. Xiè Zhàozhè rejected the popular view that 'all hares in the world are female, and only the hare in the moon is male, this being the reason that hares [in popular belief] become pregnant at the time of the full moon.' He argued that 'if you were to place some hares in a darkened room, so that for a full year they were not allowed to see the moon, would there be none of them who became pregnant?'[54] This is only a thought-experiment, and its negative answer is thought to be self-evident, but it does pose a clear question and require the creation of an 'unnatural' situation by the imposition of specified controls – the confinement of wild animals in a closed and darkened room. The hydrological model introduced in the next section also had most of the attributes of an experiment, as did some of Zhu Zàiyù's work on acoustics discussed later.

Intermediate between observation and true experiment was what can provisionally be called 'proto-experiment'. Examples can be found in the essay on *qì* (or 'pneuma') by the seventeenth-century expert on technology, Sòng Yìngxīng. He noted that to produce ceramics that resonate with a musical note, it is not adequate to dry the clay in the sun; it has to be fired in a kiln. In similar vein, he remarked that if a bell is filled with earth it will make no sound, or if a musical stone is leant against a wall, it will no longer resonate properly.[55]

All the components needed for the experimental style of thinking, plus practical expertise, were present in premodern China. They did not routinely combine in a form that embodied the full experimental style, but on occasions, both in a formal and a substantial sense, they could and did.

MODELLING

The power of models to simulate phenomena in a way that deepens understanding has become more evident in the last half-century. Computer models can replicate such complex systems as hydrological processes in river catchments, births and deaths in populations, and even the global climate, with a high fidelity to the basic patterns. Premodern models were exclusively physical, and lay along a continuum from those that reproduced only the visible *appearance* of a process to those that aimed to recreate an *actual* process on a miniature scale by the same causal mechanisms that operated in nature. In general, this latter goal is intrinsically unattainable. At some point in the building of a physical model, scale acquires an absolute rather than a relative effect. Thus, using actual sediments (gravels, sands, silts and muds) in a scaled-down physical model of an estuary yields effects uncharacteristic of the real system, and substitute 'sediments' (typically, fine-grained sawdust) have to be used instead.[56] Historical examples of models that lay near the other end of the continuum were the armillary spheres and celestial globes, driven by hydraulic clockwork, and approximately matching the rotation of the heavens. They seem to have first appeared in China in the second century CE. Needham and Wang cite Gé Hóng's description of that attributed to Zhang Héng around that period:

> [Those who] have discoursed upon the theory of the heavens [have] ... considered that in order to trace the paths and degrees of motion of the Seven Luminaries, to observe the calendrical phenomena and the times of dawn and dusk, ... to investigate the divisions of the clepsydra and to predict the lengthening and shortening of the shadow of the gnomon, [finally] verifying all these changes by phenological observations[57] – there was no instrument more precise than the [computational] armillary. [Thus] Zhang Héng made his bronze armillary sphere and set it up in a closed chamber, where it rotated by the [force of] flowing water. Then, the order having been given for the doors to be shut, the observer in charge of it would call out to the watcher on the observatory platform, saying that the sphere showed that such and such a star was just rising, or another star just culminating, or another star just setting. Everything was found to correspond [with the phenomena] like [the two halves of] a tally.[58]

The movements of the sun, moon and planets may have been represented by beads strung on threads somehow attached to the sphere or globe. This was certainly what seems to have been done later.[59] Such devices embodied the idea of modelling.

Near the other end of the continuum was the model of the Hángzhou tidal bore built around 1224 by Zhu Zhongyôu. There had been controversy about why and how this phenomenon occurred. Under the Northern Sòng dynasty, Yàn Sù had focused on the mid-bay sandbar in his *Hǎicháo [tú] lùn* [An [illustrated] theory of the tides]:

Someone may ask: 'In all the seas the full tide comes slowly. Only on the Zhè [Qiántáng] river [and Hángzhou Bay] does the wave arrive in a horizontal line like a range of mountains, as impetuous as thunder, flying crosswise between the two shores, with its snowy banks spurting forth at the side, roaring and rising up, overflowing in its haste, and hissing fearfully. Is it possible to be told the reason for its thus swelling up in anger? ...'

Under the water there is a sandbar stretching from north to south that presents an intervening obstacle to the waves, and abruptly blocks the force of the tides. ... When the moon is passing through [the phases of the cycle represented by] the trigrams *zhèn* [East] or *dùi* [West], another tide is already rising, but in the Qiántáng alone the water has still not come to a stop. When the moon passes through [the phases of the cycle represented by] the trigrams *sùn* [Southeast] or *qian* [Northwest] [hence one-eighth of a cycle later], the tide has half-arrived. The waves, turbid [with sediment], hold themselves back and become congested. More water comes from behind them, and they thereupon overflow across the sandbar with fierce anger and a sudden surge. Thus it is that it rises up and becomes the wave [of a tidal bore]. It is not a phenomenon caused by the shallowness and constriction of the river and [coastal] hills.[60]

Zhu Zhongyôu, over a century later, tells in his poem *Cháozè* [Mysteries of the tides] how he had built a model to reproduce this phenomenon:

Would you now like to look into the matter in detail? Let me try testing it with you by examining what goes on in a ditch.

If you fill a ditch with water, it will invariably enter smoothly until the ditch is half full. [Next] pile up some fragmented stones so that they form a rough and irregular surface [across it]. Let the water flow in from the upper end, and then out, so that the position-power of the water makes it pass over this roughness before it drains away. There will be nothing surprising in the turbulent flow of the water.

What Mr. Yàn called a 'bar' is sediment under the water. ... The water has to make up the difference in level before it flows on in. When the tide is increasing, but has not yet reached [the level of] the bar, the Qiántáng River [upstream] is still quite empty. After the tide has grown and confronts it head on, it discharges from the bar as though through a sluice into the river. ...

The tide is displaced and shifted by these sediments and gravels,[61] so that its turbulent flow that 'shakes the skies and makes the earth tremble' advances like a high cliff. This is the pattern-principle of water. ... That which I have called the 'rough and irregular surface' [in the ditch serving as a model] is simply like the bar.[62]

This model was wrong in its essentials. A bore is a form of shock-wave. Waves in shallow water move at a celerity (c) determined by the depth (d), namely $c = \sqrt{(gd)}$, and when a narrowing channel or steepening gradient forces the wave-front to exceed this speed, a wall of water builds up.[63] It is conceivable that Zhu may have created a hydraulic jump with his model. The most familiar version of this is when a flow from a tap hits a basin surface, and at first moves away from the point of impact in a thin layer, but then abruptly passes through a foamy discontinuity and becomes much thicker.[64] It is hard to believe that he could have created in this way what he really needed, namely a *moving* wall of water. Further, since the end of the seventeenth century no sandbar of the type described here has been observed in Hángzhōu Bay. A high coefficient of friction on an estuary bed also tends to diminish rather than enhance the likelihood of a bore forming.[65]

The conception and construction of models, both representational and replicational, are nonetheless established.

TAXONOMY

The taxonomic style of thinking goes back in Europe at least as far as Aristotle and Theophrastus.[66] It is also found in some premodern non-Western societies.[67] In contrast to the probabilistic style, it is deeply rooted in the human psyche. Differences have existed, however, in the degree of system with which it has been pursued, and the motives behind the classification. Here, as an illustration of the Chinese case, I will give an overview of the *Bêncâo gangmù* [Pharmacopoeia arranged by main headings and subheadings] published by Lî Shízhen in 1596.[68] This summary suggests that, since it was not markedly 'inferior' to most of its European near-contemporaries, the problem to be solved is why this work had no successors that structurally improved on it.

Taxonomy is the conceptual arrangement of all the individual members of a large class of entities or phenomena possessing an overall similarity, but differences in detail, into a structured pattern on the basis of explicitly defined characteristics. This pattern is commonly a hierarchy in which members at each level are nested in a unique place within one member of the level above. Thus for us today the species of plants are members of genera, and the genera of families, the families of orders, the orders of classes, and the classes of divisions or phyla.

A taxonomic system attempts to capture in words, in an economical manner, those aspects of a coherent subset of the observable world (such as fishes, diseases or languages) that the taxonomist decides are relevant to his or her purpose. Different purposes inspire differing structures. A field-guide designed for the easy identification of a limited number of plants is unlikely to use markers for differentiation identical with those of more strictly scientific systems concerned with mapping shared descent from assumed, if perhaps unidentifiable, common ancestors. A taxonomic structure is thus in one sense 'artificial'. Among other desiderata it has to ignore all observable features in the material that are not relevant to its concerns (for example, size and colour – in general – for an evolutionary structure). It would be of little practical use, however, if it did not also capture aspects of the 'real' universe. Hence it is also 'natural'.

A taxonomic structure requires entities that are separable, not a continuum of forms. It also needs a vocabulary of descriptive terms free of ambiguities at the margins. Equally important, descriptions should be as little as possible distorted by implicit theoretical bias. It needs to be free of internal inconsistencies so that the assignation of an entity to a place in the structure is unique. Also stable, so that the incorporation of new knowledge disrupts it as little as possible. Membership of an item in one or other of two otherwise equally possible sets should be determined by a single criterion (that is, it should be 'monothetic'), rather than by the possession of only some of a number of a specific group of characteristics ('polythetic' determination). In practice, these requirements are rarely fully realized in systems of actual use and are capable of interacting with information drawn from outside the systems themselves.[69]

A European example of how a division of the natural world was brought into an orderly arrangement by applying a coherent set of criteria is that of Andrea Cesalpino in the late sixteenth century for flowering plants,[70] later fully developed by Linnaeus in the middle of the ~~seven~~teenth *l8* (though abandoned by plant taxonomists in the second quarter of the nineteenth). This was classification according to the differences in reproductive systems. In the Linnaean version, the 'classes' were determined by the number and relative lengths of the male organs, that is, the stamens. The classes were subdivided into 'orders' defined according to the numbers of the female organs, that is styles/stigmas, with distinctions between cases where stamens and pistils did or did not occur in the same flower, and where male and female flowers were on different plants.[71]

If we contrast this with the ten 'categories' (*lèi*) for the 610 entries in the 'division' of medicinal plants (*câo*) in Lî Shízhen's *Pharmacopoeia* the difference in approach is evident:

1. Mountain plants (*shancâo*)
2. Oderiferous plants (*fangcâo*)
3. Wetland plants (*xícao*)
4. Poisonous plants (*dúcâo*)
5. Creeping plants (*màncâo*)
6. Water-plants (*shûicâo*)
7. Rock-plants (*shícâo*)
8. Mosses (*tái*)
9. Miscellaneous (*zácâo*)
10. Plants with reputation but no use (*yôumíng wèiyòng*)

The criteria in this list are disparate. The first, third, sixth and seventh categories use habitat. The second and fourth use chemical properties. The fifth follows external form, and the eighth, mosses, are of course today in a separate phylum, the Bryophyta, defined by the lack of vascular tissue. The ninth category represents an abandonment of the attempt at classification. The tenth draws our attention back to the medical purpose of Lî's work.

The top-level divisional structure of the *Pharmacopoeia* is arranged in a sequence that reflects an ascending order of metaphysical importance, beginning with forms of matter-energy (*qì*) that have not congealed into a definite shape, and ending with medicines derived from the bodies of human beings (hair, nails, etc.). The 16 divisions (*bù*) are listed below, followed by the number of categories (*lèi*) in each, and the total number of 'types' (*zhông*):

1. Waters :: 2 : 43
2. Fires :: 1 : 11
3. Earths :: 1 : 61
4. Metals and rocks :: 4 : 160
5. [Medicinal] plants :: 10 : 610
6. Grains :: 4: 73
7. Vegetables :: 5 : 105
8. Fruits :: 6 : 127
9. Trees :: 6 : 180
10. [Medically useful] manufactured products and devices :: 2 : 79

11. 'Beasties' :: 3 : 106
12. Scaly creatures :: 4 : 94
13. Shelly creatures :: 2 : 46
14. Birds :: 4 : 77
15. Hairy quadrupeds :: 5 : 86
16. [Medically useful products from the bodies of] human beings :: 1 : 35

Within a few of the divisions there are attempts to use a single criterion systematically. Thus the three categories for the eleventh division, 'beasties' (*chóng*), are: 11.1 Born from eggs; 11.2 Born from transformations; and 11.3 Born from moisture. Types in the first category include bees, wasps, butterflies, cicadas, dragonflies, ants, spiders, lice, the silkworm moth, and scorpions (which last are in fact born live). Those in the second category include maggots, beetles including the dung-beetle, locusts, the mole-cricket, another type of cicada, and the firefly. Typical members of the third category are frogs and toads, snails, slugs and centipedes. Regardless of its inaccuracies, this is potentially the beginning of a more developed taxonomic style of thinking.

Li's comments show him wrestling with systematizing such diversity:

> Beasties are the smallest of living creatures. To an extreme degree their category [*lèi*] contains entities of different sorts (*fán*). Therefore their entries are grouped under the written character composed of the thrice repeated graph for 'insect' [*chóng*] so as to conform with this idea.
>
> *Note:* According to the *Kǎogong jì* [The *Record of the Scrutiny of Crafts*, a work of the fifth century BCE that was incorporated into the *Rituals of the Zhou*]: The class [*shū*] we designate as 'small beasties' has members that may have external or internal bones; move upside-down [like a spider on a web[72]], sideways [like a crab], in zigzags, or in a continuous line; and that give voice through their throats, beaks, flanks, wings, abdomens, or chests. Although they are minute entities, and not on a par with the unicorn, phoenix, turtle, or dragon, they even so have forms with feathers, fur, scales, or shells, or else are naked. They differ in how they give birth, whether from the womb, or from eggs, or from the transformations of wind and moisture. [Even] their wrigglings contain a magical force [*líng*], each endowed with its own character and energy-vitality [*qì*]. One records their useful contributions and explains their poisons. The sage therefore discriminates between them.[73]

The taxonomic impulse is undeniable.

A substantial part of the material was explicitly pasted in from earlier sources, perhaps for reference purposes, since it seems unlikely he had personally validated all the information. (The book is, after all, enormous.) For example, in a sub-entry under 'cinnabar' (*zhusha*, i.e. HgS), he writes

> When someone has already died from the spasms of the sinews created by quasi-cholera [*huòluàn*], but there is still a slight warmth below the heart, two ounces of cinnabar mixed with three of beeswax and warmed until it is emitting fumes, will cause the sweat to pour out, and *he will return to life*.[74]

Another entry that strains credulity is that on the 'Blue Bug' (*qingfú*, a term popularly used to refer to copper cash). Li notes six alternative names for this insect and a seventh possible identification, and gives a *scholarly* presentation of the various texts in which it is mentioned. Here is a sample:

> The *Cángqì* says: The Blue Bug lives in the far south. Its shape is like that of a cicada. Its juveniles [larvae, or nymphs is not clear] adhere to a tree. If you get hold of some and rub them on some copper coins, *all the coins will come back to where they started*. The *Soushénjì* says: ... Its form is about the same size as a cicada's. It is bitter but pleasant to eat. The juveniles adhere to blades of grass like the eggs of the silk moth. If you get hold of some juveniles then the mother will come flying to you. If you kill the mother and smear her on some coins, and smear the juveniles on the cord that threads through [a thousand of] them, *you may spend the cash but they will come back to you of their own accord*. ... Lî Xún says: According to the *Record of Strange Things* ... they live in the mountains of south China. Males and females constantly reside in the same place and never abandon each other. They are the hue of blue metal. ... *They can also cause retention of the semen and the cessation of the urine*. ...
> Lî Shízhen says: ... All these statements resemble one another. Only the *Cángqì* is slightly different in saying that the juveniles adhere to a tree. However Mr. Xû says in his dictionary, the *Written Speech* [the famous *Shuowén*] that 'the Blue Bug is an aquatic beast.' The explanation is that it is aquatic but gives birth to its young in trees.

He observes that another account on south China refers to a blue cicada-like insect, the *pángjiàng*, used to make an aphrodisiac. Since this fits well with what Lî Xún says about the males and females never leaving each other and the retention of the semen, he concludes 'I suspect that this is a type of Blue Bug'. It seems certain that Lî Shízhen had never himself seen a Blue Bug, but he still lists its properties, such as 'removing cold aether, and causing people to shine with well-being,' and 'retaining the semen and causing the withholding of the urine'. For each of these assertions he gives footnotes to the authorities, but does not quite list himself, as he often does, among them.[75]

In other words, the book has two dimensions. In part practical, it is also an anthology that retransmits earlier statements: 'scholarly' but not scientific. Li does not feel it essential in such cases to refer to recent validation or invalidation by experience, though on occasion he does so. We have earlier noted his verification, by dissection, of the belief that pangolins eat ants.

The *Bêncâo gangmù* contains a semi-systematic four- to five-level filing structure for useful information about diseases and *materia medica*. This appears in two chapters on 'The drugs for the principal treatments of specific diseases' (*bâibìng zhûzhì yào*). These link the disease entity diagnosed by the practitioner with the appropriate medicines. They are thus organized roughly *in reverse fashion* from the main body of the pharmacopoeia which gives the leading place to the *materia medica* and then lists its uses. The top-level entries in these two chapters are groups of diseases or medical conditions conceived as being related. The keys to these entries are defined in terms of different types of criteria: location (such as 'eyes'); symptoms (such as 'inability to keep food down' or 'prolapse of the anus'); syndromes defining disease entities (like 'jaundice' and 'beriberi'); aetiology (diseases caused by 'aethers', for example, or 'afflictions caused by vacuities' (*xusûn*) or possession by 'evil spirits', as well as injuries caused by weapons and wild animals); epidemiology (seasonal and other 'epidemics' being the most obvious); and diseases and disorders associated with a particular phase of life (notably pregnancy and childbirth, and the illnesses of neonates).

The second level of subdivision separates these groups according to theoretical concepts used in traditional medical theory. These include such categories as 'Humidity', 'Warmth', 'Vacuity', and the *yin* and the *yáng* forces. At or near the end of such subdivisions there is also often the important category of 'Accumulation-Congestion' (*jizhì*) , which can be difficult to handle conceptually as it was also thought to manifest itself as an excessive flow – for example diarrhoea – due to prior accumulation. The first two levels can be illustrated by the simple case of quasi-cholera:

A. Quasi-cholera [*huòluàn*]: There are (i) humid-hot types, and (ii) cool-humid types. In both the Seven Passions [*qiqíng*] are damaged internally and the Six Agents [*liùqì*] affected externally.

There follow three sections: 'Humid-Hot', 'Cool-Humid' and 'Accumulation-Congestion'.

Within each of these sections suitable medicines are grouped according to the domains from which they come, such as Plants, Grains, Fruits, Waters and Soils, Metals and Rocks, Beasties and Animals, and Human Beings. This constitutes the third level. Within each such domain the relevant individual medicines are listed, or small groups of medicines that have to be used together synergistically. Often a note will give the effects of a medicine on a symptom, and prescribe the preparation and use. An illustration, continuing the case just given, would be:

B.Quasi-cholera/Humid-Hot type/Plant medicines/*Elsholtzia ciliata* [*xiangru*, formerly *E. crissata*, a member of the Labiatae]. When quasi-cholera causes spasms of the sinews and the stomach hurts, boil *Elsholtzia ciliata* in water and administer the liquid orally.

An optional fifth level of classification specified the medicine to be used when particular symptoms were targeted, or the patient was in a particular category. Thus:

> C. Quasi-cholera/Humid-Hot type/Metals and Rocks. If *a small child* is afflicted by heat, and his vomit and diarrhoea have *a yellow colour*: Gypsum [*shígao*]. Grind some gypsum to a powder with cold water and solidified *Glycyrrhiza uralensis* [*gancâo*, liquorice], and have him take it orally.[76]

Diagnosis is taken for granted. This is not a defect in the system as its purpose is to guide the prescription once the diagnosis has been made, not to steer the identification of the disease. A lot of diagnostic information is incorporated in the entries, typically in the form in B above. But it is not part of the structure, only of the contents.

In these two chapters we thus have an approximation to a tree for making decisions, *given* identification, rather than the more familiar form leading *to* identification via a sequence of ordered observable criteria. It is an approximation because in most cases there is a range in the choice of medicines. Only at the fifth level does there tend to be a unique or near-unique answer.

The same basic ingredient might also be prepared differently for different diseases or 'sub-diseases' (when the fifth level was used). An example is *huánglián*, which is *Coptis chinensis*, and also *C. deltoidea*, a bitter-tasting member of the Ranunculaceae. For diarrhoea of the humid-warm type it was taken with powdered raw ginger, but if there was an 'accumulation' of food causing 'stomach diarrhoea', it was taken in pill form made with onion.[77] In the case of dysentery of the humid-warm type there were nine sub-diseases for all of which the mode of preparation of *huánglián* differed. A single example by way of illustration is:

> For hot-poisonous red dysentery, boil it in water, and leave it exposed to the air overnight, then administer warm orally. For small children add honey, or else roast it, add powdered *Angelica sinensis* [*danggui*], musk from the deer *Moschus moschiferus* [*shèxiang*], and administer orally in rice-water.[78]

It would need study to determine the extent to which the same medicine was used prepared in the same way for different diseases. Some of the basic remedies, like ginseng, certainly were.

In the sense that a taxonomy is an ordering of a conceptually coherent set of distinct entities that share the majority of their properties but nonetheless differ critically in at least one, this system ranks as a taxonomy. That it was an imperfect one is evident. Many of the criteria depended on the allocation of entities to conceptual categories for which there were no operationally defined observational tests. Examples are notions like 'yin' and 'yáng', and 'aether' (*qi*) and 'wind' (*feng*) in a somatic sense, as well as such either-or notions as 'warm' and 'humid' when applied to diseases.

The late-imperial pharmacological tradition produced substantive works after Lî Shízhen's *Bêncâo gangmù*. Besides correcting errors and clarifying older materials, they recorded the arrival of new plants such as tobacco and tapioca, and included both new remedies and new categories, flowers being an example, that were not included in the *Bêncâo gangmù*. It does not seem, though, that there were any significant advances of a taxonomic nature.[79] The problem, yet again, is why further development stagnated.

PROBABILISTIC THINKING

Probabilistic thinking deals with repeating patterns that can be expressed numerically, and where the causes of the particular events that constitute the patterns cannot be determined. Premodern China produced many such patterns, including the use of oracles such as the yarrow-stalk casting and coin-tossing of the *Book of Changes*, dice-throwing and other forms of gaming, central and local governmental records including not only taxes and landholdings, but also such items as the ages at marriage, bereavement, and death of faithful widows receiving official recognition, and the frequencies of natural disasters like droughts and floods. This style of thinking can be subdivided into analyses of *distributions*, or 'statistics', and of *pathways*, that is the *distinct* ways in which a given outcome can occur, such as getting a total score of 11 when throwing two dice.[80]

Statistics
From early-imperial times the Chinese could calculate an arithmetical mean. It can be found as a by-product of the operation called 'equalizing allocations' (*píngfēn*). It is appropriate to call this an 'arithmetical' as opposed to a 'statistical' average in that it was not thought of as a measure of central tendency. 'Equalizing allocations' *removed* the pattern of a distribution rather than characterizing it. Thus problem 16 in the first chapter of the *Nine Chapters on the Mathematical Art* from the third century CE reads:[81]

> Again, we take as given 1 part out of a division into 2, 2 parts out of a division into 3, and 3 parts out of a division into 4. We ask: if we reduce the excessive and augment the insufficient, what is the quantity for each of these that will bring it into equality (*píng*) with the others [similarly treated]?[82]

This requires finding the mean of $1/2$, $2/3$ and $3/4$ as an intermediate step. The answer given is: 'Reduce the 2 parts of a division into 3 by 1, and the 3 parts of a division into 4 by 4, and add [the sum of] these to the 1 part of a division into 2, after which all of them will be equal at 23 parts of a division into 36.' We would understand this as

$$\frac{18 + 5}{36} = \frac{24 - 1}{36} = \frac{27 - 4}{36} = \frac{23}{36}$$

How the *Nine Chapters* reached its result may be summarized as follows: (1) The three fractions were added as we do today. That is, each dividend was multiplied by the denominators of the other fractions, these results summed, and then the sum divided by the product of all the denominators. This gives 46/24. The mean was derived by dividing this by the number of items, giving 46/72. This is not stated; perhaps it was thought obvious. (2) Each of the three products of a dividend with the other denominators (that is, 12, 16 and 18) was multiplied by the number of items, here 3. The same was done for the product of the denominators (24). The *differences* of each fraction from the mean were then determined, giving

$$\frac{36 - 46}{72} = \frac{-10}{72}, \frac{48 - 46}{72} = \frac{+2}{72}, \text{ and } \frac{54 - 46}{72} = \frac{+8}{72}$$

We would think of this today as calculating $x_i - x$. (3) The signs of the differences were then reversed in order to 'reduce the excessive and augment the insufficient,' and the expressions simplified to the lowest ratios of integrals. In the third chapter *weights* are also used in redistributions to create the sort of 'equity' (not equality) that allows for differences in such characteristics as status or, for animals, species.

The problems derived from the need of bureaucrats to make the burden of taxes and labour-services 'equitable' (*jun*) seem at first sight to provide cases of statistical distributions. There is an example in the *Nine Chapters*,[83] where four 'counties' have to share equally the burden of moving 250,000 'bushels' of grain to a given place using 10,000 carts.[84] The populations and the required travel-times are shown in Table 1.

Table 1 *The equitable allocation of ten thousand transport carts*

County	Households	Days travel	h/d	$(h/d)/\sum(h/d)$		Carts (rounded)
A	10,000	8	1250	0.3324	47	3324
B	9,500	10	950	0.2526	60	2527
C	12,350	13	950	0.2526	60	2527
D	12,200	20	610	0.1622	34	1622
Totals			3760	1.0000	01	10,000

In what numbers should carts be assigned to each county?

We would probably solve this by assuming that each cart carries 25 bushels and makes one complete single trip. The fourth column, h/d, shows the number of households available to provide the support for one cart-day of one complete multi-day trip of the time required for each county. The fifth column shows the proportion of the total of all the entries in the fourth column that is available from each county. Since the more households available per cart-day the lighter the burden per household, this is counterbalanced by assigning *pro rata* more of a notional single cart to the counties with more disposable resources. Since 10,000 carts are available, the final column is derived by multiplying the entries in the preceding column by this number, and then rounding. The burden of cart-days per household, cd/h, is now close to about 2.7 for all four counties. (Thus for A, $(3324) \times (8/10,000) = 2.66$, etc.)

The answer given in the *Nine Chapters* is the same, and the method similar. That is, it calculates the h/d values and uses the ratio of each such value to the sum of the values as a set of weights, *which sum to unity*, though it does not calculate cd/h. What is striking is that the operation is aimed at *neutralizing* the effects of the unequal distributions of populations and travel-times. The concept of 'equity' is also not the same as that of 'average'. This kind of dataset would appear to us to lend itself to further – statistical – questions, but they do not seem to have been asked.

Pathway Analysis

Indirect evidence suggests a deeper grasp of probability pathways than is found in explicit form in any text I have so far seen.

The Chinese scripture known in the West as *The Book of Changes*, and in China as the *Yì* [Changes] or *Zhou Yì* [Changes of the Zhou], offers a striking illustration.[85] Its mechanisms were familiar over twenty-odd centuries to hundreds of millions of Chinese since its use became well established in the third century BCE.[86]

The book is based on 64 'hexagrams' (*guà*). In their classical form these are sets of 6 horizontal parallel lines constituting the 2^6 possible distinct orderings of 2 types of line, which can be characterized as complete '—' and broken '– –', or as 'hard' and 'soft', or as 'yáng' and 'yin'. Thus the idea of a *permutation* is implicit in the evolved system, although the graphical origins of the hexagrams may have been sequences of archaic numerals with quite other numerological implications.[87] The underlying idea is that there is always a hexagram that either 'is', or symbolizes, any particular situation in the world as it currently applies to the questioner seeking foreknowledge and advice. The system is the abstract form of the cosmos as it moves through time.

The lines have a further property. They can be either changing (the so-called 'old' lines) or unchanging (the so-called 'young' lines). It was this innovation, associated with the Zhou dynasty, that may have given the *Changes* its name, in contrast with analogous earlier systems.[88] If the hexagram determined as the one relevant to the inquirer's question contains some lines that are changing lines, the future development of the situation is regarded as symbolized by the transition from this hexagram to the one formed by altering the changing lines into their opposites: complete to broken and vice versa.

Old notation represented changing *yáng* as 'ø' and changing *yin* as '— × —'. A variety of representations of hexagrams in modern notation is also possible. If *yáng* is 'positive', and *yin* 'negative', youth 'imaginary' and old age 'real', we could for example set old *yáng* = 1, young *yin* = i, old *yin* = -1, and young *yáng* = $-i$, where i is the $\sqrt{-1}$. Representing inherent change would require multiplying each *changing* numeral in the string of 6 by i, so old *yin* becomes young *yáng*, and old *yáng* young *yin*. One justification for these assignations of values is that the traditional identification of the 4 types of line with the 4 seasons of the year can be made to match up. Old yáng is summer, young yin autumn, old yin winter, and young yáng spring.[89] So, starting in summer, successive multiplications by i cycle around the years: $< 1, i, -1, -i, 1, \text{etc.} >$. The traditional system does not incorporate any *formal* young-yang-to-old-yang or young-yin-to-old-yin transitions but it is implicitly cyclical.

The simplest way of determining the applicable hexagram was by 6 tosses of 3 coins at time, each 3-coin toss determining a line, beginning at the bottom. Traditional Chinese coins had an obverse and reverse, like our 'heads' and 'tails'. We may say that **HHH** yielded a *changing* complete line $< 1 >$, TTT a changing broken one $< -1 >$, the 3 permutations of TTH an *unchanging* complete line $< -i >$, and the 3 of HHT an unchanging broken one $< i >$.

The coin oracle is symmetrical as regards the probabilities for *yin* and *yáng*, and does not yield information useful for our present purposes. The most highly regarded method for determining the appropriate hexagram was, however, the manipulation of the stalks of the common yarrow (*Achillea millefolia*),[90] according to an involved procedure.[91] As I show below, *the yarrow-stalk oracle did not give precisely the same frequencies as the coin oracle*. These discrepancies apparently did not prompt an inquiry into the reasons behind them.

Broadly speaking, each of the 6 lines was determined one at a time by a 3-step procedure that left a certain number of stalks in 3 groups in the hand of the diviner. These triples of numbers determined the nature of the line, as shown in Table 2.[92] The values assigned to the various sets of stalks had a metaphysical rationale, but can be treated here as simply conventional and, from a mathematical point of view, arbitrary. Finding the probabilities of these triples, such as $(5,4,4) = < +1 >$, or $(9,8,8) = < -1 >$, requires a description of the essential parts of the divining procedure constituting the 3 steps in Table 3. Since this is tedious, those interested are referred to Appendix B of my paper on deposit in the Needham Research Institute in Cambridge. Table 3 shows the probabilities for the step entries in Table 2.

Table 2 *Lines determined by triples of stalks left in the hand of the diviner*

	Stalks			Modern symbol	Traditional name
Step	**1**	**2**	**3**		
	5	4	4	$+1$	Old Yáng
	9	8	8	-1	Old Yin
	9	8	4	$-i$	Young Yáng
	5	8	8	$-i$	Young Yáng
	9	4	8	$-i$	Young Yáng
	9	4	4	$+i$	Young Yin
	5	4	8	$+i$	Young Yin
	5	8	4	$+i$	Young Yin

From Table 3 it seems likely that when the changing stalks were introduced into divination the yarrow-stalk system was *engineered*, either with analytical *understanding* or by empirical *experiment*, to produce an equal overall number of *yáng* and *yin* outcomes. The discrepancy in the two probabilities in the *yáng* and *yin* rows is only just over 19 hundred-thousandths. The structure in Table 3 is, however, far from intuitively self-evident. Twelve separate components go to make up each of the *yáng* and the *yin* probabilities. In other words, unlike the coin oracle, it could hardly have been set up on simple first principles to give this result.

It follows that either there was a substantial understanding of probability, which was kept *secret*, among those who created and refined the system,[93] or there was, at least in private and during a certain period, *a minimally reverential experimental attitude* to its mechanism, which was adjusted till it gave the best observed results. I favour the second alternative; the first cannot be ruled out.

Table 3   *Probabilities for the components determining the lines shown in Table 2*

	Stalks			Symbol	Probability
Step	**1**	**2**	**3**		
	36/47	22/42	20/38	+1	0.211 166 213
	11/47	18/38	14/30	−1	0.051 735 722
	11/47	18/38	16/30	−*i*	0.059 126 540
	36/47	20/42	16/34	−*i*	0.171 643 125
	11/47	20/38	16/34	−*i*	0.057 967 196
	11/47	20/38	18/34	+*i*	0.065 213 095
	36/47	22/42	18/38	+*i*	0.190 049 592
	36/47	20/42	18/34	+*i*	0.193 098 516
Yáng	(rows 1, 3, 4, 5)				0.499 903 074
Yin	(rows 2, 6, 7, 8)				0.500 096 925
Total					0.999 999 999

The consequence of giving primacy to overall *yáng/yin* parity was *disequilibrium* between them in subsidiary respects. The yarrow-stalk system was more than 4 times more likely to yield a changing *yáng* < +1 > than a changing *yin* < −1 >. A young *yin* (the sum of the three < +*i* >s) was more than one-and-a-half times more likely than a young *yáng* (the sum of the three < −*i* >s). What logic decreed that the universe was so lopsided? The analyses of Shào Yong (in the Sòng) and other *Changes* numerologists constantly implies balance between the complementary cosmic forces. The centre of the philosophical apple had a rotten mathematical core, whether unrecognized or unacknowledged we cannot for the moment tell.

The probability of getting a changing line, (p< +1 > + p< −1 >), was also 0.263 as opposed to the 0.25 when tossing coins. Some experienced gamblers are reputed to be able to notice differences of this order. It passed without comment.

The foregoing refers to antiquity. For the middle ages circumstantial evidence points to some persons – such as the operators of entertainments based on probabilities, and shopkeepers who sold through a gaming mechanism – having had a familiarity with the approximate odds involved, if only for economic survival.

The multiplication rule was implicit in some traditional spill-, dice-, and coin-tossing games using multiple coins, etc. Bets were made in two main ways: (1) on exactly how many heads or tails would be showing, and (2) on which pattern would appear. For example, one possibility was that the coins would be either all heads or all tails, or else that, with an even number of coins, there would be exactly half of each. For 3 coins the outcomes, ignoring different orderings, would tend to the limits:

$$
\begin{array}{llll}
\text{HHH} & (1/2)^3 & & = & 1/8 \\
\text{THH} & (1/2)^3 \times 3!/2! & & = & 3/8 \\
\text{TTH} & (1/2)^3 \times 3!/2! & & = & 3/8 \\
\text{TTT} & (1/2)^3 & & = & 1/8 \\
\end{array}
$$

where the '$n!$' indicates the factorial $1 \times 2 \times 3 \times \ldots n$. The numerators of the fractions (1, 3, 3, 1) can be found from Pascal's triangle.[94] Chinese mathematicians had been familiar with this array since at least Sòng times, but do not seem to have used it to analyse probability.[95] For patterns, using the examples given above, and assuming 4 coins, HHHH or TTTT would have a frequency tending toward $(1/2)^4 + (1/2)^4 = 2/16$, and HHTT, which has 6 possible orderings, would have a frequency tending toward

$$
(1/2)^4 \times \frac{4!}{2! \times 2!} = 6/16
$$

We have no records telling us how payoffs were related to frequencies. There are references, however, to Sòng dynasty shopkeepers using coin-tossing to boost demand for their wares. If the customer won a game with the shop-owner, he got back his stake and also the goods free of charge. If he lost, the shopkeeper kept the stake, and the customer departed with no goods. The term for this practice was 'sale by gambling' (*pumài*, or *bómài*). Only shopkeepers who had a good notion of the odds would have stayed long in business.[96] But there is no reason to believe that they knew how these odds could be calculated theoretically.

Mèng Yuánlâo's *Dream of the Glories of the Eastern Capital*, which describes twelfth-century Kaifeng, has this to say of gambling-purchase:[97]

> On the first day of the first month of the lunar calendar, the prefectural government of Kaifeng permitted three days of purchase by gambling (*guanpu*). ... In the city quarters and blocks of dwellings people would chant out that such goods as foodstuffs, tools, fruits, firewood, and coal were available for *purchase by gambling*. Multi-coloured covered stalls were put up side by side. ... Here were spread out on display headgear, combs, pearls, jades, ornaments for

the head, clothing, artificial flowers, curios and playthings. Dance-floors were laid out between them and also halls for singers, while carriages and horses chased after each other in criss-crossing fashion. Towards evening the women from families of high social standing enjoyed gambling-purchasing (*guanbó*) in an uninhibited manner, joined the audiences at entertainments, and drank and feasted in the restaurants in the markets. This had become so customary that no one tittered with surprise at seeing it. Three days of gambling-purchasing were likewise permitted at the Cold Food Festival in the spring and at the winter solstice.

According to the eighteenth-century writer Lǐ Dôu, there were no restrictions on time and place for the practice under the Southern Sòng.[98] Hóng Mài, writing under this dynasty, thus tells us of a certain Lǐ Jiàngshì, that[99]

> It so happened that vendors carrying Yôngjia yellow oranges went past the door, and he shouted out in a lively voice that he would gamble for them. After losing ten thousand cash, the irritation showed on his face. 'I have lost ten thousand cash,' he said, 'without a single orange passing my lips.'

'Ten thousand' here presumably means only 'far too much'.

A hint of the methods used can be gleaned from a tale told by Zhang Zhongwén relating to the year 1217.[100] It tells how Zhèng Fùlî, a medical specialist who was one of the followers of Kông Wèi, the magistrate of Gaoan in present-day Jiangxi province, abducted his patron's daughter. The note in parentheses is in the original.

> Several days before this happened, Zhèng saw people in the [provisional] capital at Hángzhou gambling-buying chickens. (The rate at the [provisional] capital was that 3 copper cash were thrown, 10 *chún* obtaining the chicken and the 3-cash stake back.) This prompted him to gamble to decide whether or not he would do it. He prayed to the dicing coins (*tóuqián*) that he would obtain a *chún chéng*, wishing to abduct this girl. He forthwith obtained a *chún zi*. Putting his reliance on prayer yet again, he gambled a second time, getting another *chún*. He thereupon followed his reckless lusts, and, taking advantage of Mr. Kông having to offer a sacrifice at the Yâzhai Altar, abducted her that night.

The most plausible *mathematical* interpretation of the technical terms is that the 3 coins constituting the stake for each attempt were thrown 10 times, and to win they had to show a 'pure pair' each time, that is, no triples. The Kangxi and Couvreur's dictionaries give a rare meaning of *chún*, pronounced *qún*, as 'a pair' of calculating rods, so this is a tenable lexical reading. The probability of no triples in 10 throws of 3 coins is $0.75^{10} = 0.056$.

Could the seller have had some easy way of at least *estimating* the probabilities? Provided that either calculation or experience showed him that, on average, for each throw of the 3 coins, one quarter of those doing the throwing lost and were eliminated (HHH and TTT being together one quarter of the 8 possible outcomes), it would not have been hard to see that if, say, 64 people started throwing together, after one throw of 3 coins each, 48 of them would be left, after 2 throws, 36, and after 3 throws only 27, etc. This method of deducting a quarter 10 times would have indicated that, approximately, 3.6 out of the 64 starters would finally win a chicken and their stake back. The *lexically* plausible alternative interpretation of *chún* as 'pure' *triples* of either 'heads' or 'tails', allowing either of the latter to count on any given throw, gives a probability of success of 0.25^{10}, or close to 1 in a million, which is not a viable commercial arrangement.

Finally, we know how the ancient game of Lay Out The Coins (*tanqián*)[101] was played when adopted in the later nineteenth century as the basis for the *fantan* in state-supported dens to provide finance for the navy.[102] An arbitrary quantity of copper coins was drawn at random from a heap called 'the cash surface' (*qiánpí*), and put into a receptacle (the *tanzhong*). The participants then laid bets on the residue that would remain when the contents of the receptacle were later counted off by fours. (Thus 30, which is $(4 \times 7) + 2$, would leave 2.) The 4 possibilities, or 'gates' (*mén*): 1, 2, 3, and 4 or 0, were taken as a cycle for purposes of ordering. There were four types of wager:[103]

1. *Fan*. Betting on a *single* number. The payoff for success was the stake back plus 3 times the stake.
2. *Niân* or *rên*. Betting on 2 *differentiated adjacent* numbers, one of which was the *principal* and the other the *support*. If the winning number was the principal, the payoff was the stake back plus 2 times the stake. If the support was the number that came up, the stake was returned.
3. *Jiâo*. Betting on 2 *adjacent* numbers *treated equally*. If either was the number that came up, the payoff was the stake back plus 1 times the stake.
4. *Zhèng*. Betting on 3 *sequential* numbers, the one in the *middle* being the principal. The payoff for the principal was the stake back plus 1 times the stake. If either of the other 2 numbers came up, the stake was returned.

The gambling establishment derived its income from a 10% levy on winnings. This apart, is it a fair game? It seems so. Writing '*s*' for 'stake', we have the following expectations for the 4 options:

1. $(1/4) \times 4s = 1s$
2. $(1/4 \times 3s) + (1/4 \times 1s) = 1s$
3. $(1/4 \times 2s) + (1/4 \times 2s) = 1s$
4. $(1/4 \times 2s) + (1/4 \times 1s) + (1/4 \times 1s) = 1s$

The first option, *fan*, is obviously fair. In the case of the other 3 options, it is necessary to consider the effect of bets excluded by the rules governing the relative positions of the numbers chosen. Putting the principal number in bold type, where relevant, we have:

2. Permitted: **1**2, 12; **2**3, 23; **3**4, 34; **4**1, 41. *Not permitted:* 13,13, 24, 24.[104]
3. Permitted: 12, 23, 34, 41. *Not permitted:* 13, 24.
4. Permitted: 12**3**; 23**4**; 3**4**1, **4**12. *Not permitted:* **1**23, 123; **2**34, 234; **3**41, 341; **4**12, 412

In options 2 and 3, only two-thirds of the possible bets are allowed, and in option 4, only one-third. Computer simulation and analysis both show the debarred options do not affect fairness. The number of coins drawn can have a slight effect on the frequency of the 4 moduli, but simulation also shows this can be eliminated by requiring the number of cash drawn to fall within a reasonable range. In 10^7 trials for the range 31 to 98, all 4 moduli appeared at a frequency of 2.50×10^6. Analytically, the irrelevance of the restrictions to the outcomes is suggested by the observation that, given the payoffs tabulated above, a gambler limited to betting on a unique number or unique pair or triple of numbers still breaks even so long as they occur with a fair frequency.

It may have been enough for the construction of this game to see that, given numbers occurring with equal probability, the better's net gains and losses sum to zero for each of the four betting patterns. These may be expressed as the four cyclical groups respectively formed from $(-1, -1, -1, +3)$, $(-1, -1, 0, +2)$ *and* $(-1, -1, +2, 0)$, $(-1, -1, +1, +1)$, and $(-1, 0, +1, 0)$. The game thus embodies in *implicit* form four of the principles needed for an elementary calculus of probabilities: (1) The *equal probability* of a number of outcomes; (2) The *addition* of *alternative* probabilities; (3) The *exhaustive* consideration of all the possibilities; and (4) The determination of the *expectation* as the *product of the probability and the payoff*. The multiplication of probabilities is missing. Since Lay Out The Coins was an old Chinese game, direct foreign input appears unlikely. As we only have the crucial detailed information of its workings for the end of the imperial period, we have also to bear in mind the possibility that it was refined around this time, conceivably under outside influence.

There was a closer approach to probabilistic thinking in premodern China than appears explicitly in the historical record. The main reason for the silence on the pragmatic details may have been desire for secrecy on the part of the professionals who in early times created yarrow-stalk divination or later ran gaming operations. The lack of interest on the part of the scholars who wrote the records must also have played a part.

HISTORICAL DERIVATION

The presence of the concept of historical derivation can be shown by a summary of Gù Yánwǔ's *Yìnlùn* [Theory of reading pronunciations], with an illustration of his method drawn from his *Tángyùn zhèng* [Táng reading pronunciations corrected]. Both works were composed in the middle of the seventeenth century.[105] The *Yìnlùn* incorporates some of the ideas of Gù's predecessor Chén Dì (1541–1617), though Gù seems to have gone beyond Chén in the scope of his researches.

Archaic Chinese was basically a monosyllabic language. Its script, each item of which in principle represented a syllable or, normally, morpheme, was not phonetic, though a large percentage of characters from archaic times incorporated phonetic elements. Those that do are known as *xiésheng*, *xié* meaning 'in harmony' and *sheng* meaning 'sound'. The sets of 'characters in harmony' form series of words that had closely similar pronunciations about two-and-a-half millennia ago. This is an aid to reconstructing archaic pronunciations (given here according to Karlgren's *Grammata Serica Recensa*). These are marked here by an asterisk *, and simplified, since we are not concerned with the minutiae of phonology. Reconstructed Middle Chinese (Táng-period) pronunciations are marked with a dagger†.[106]

It had been apparent for a millennium when Gù wrote that the archaic and middle and modern pronunciations of a given graph were often no longer the same. Thus in the *Shi* [Scripture of the Songs] the finals of words at the end of lines that were supposed to rhyme often no longer did so. The character today pronounced *nán* 'south' is found in the *Songs* rhyming with *xin* 'heart'. This is explicable by the likelihood that they were originally pronounced something like *nem* and *syem*.[107] The assignation of words to rhyme-classes had also become internally self-contradictory as scholars had, in various periods, tried to adjust these assignations to the pronunciations of their own times or to assumed ancient values. An example, discussed by Gù, was that the three words *dong* 'east', *dong* 'winter' and *zhong* 'wine-vessel', which rhymed in modern times, had originally been placed in different categories: (*tung*, †*tung*), (*tông*, †*tuong*), and (*tyung*, †*tsywong*) respectively.[108] The problem was how to reconstruct the processes by which the archaic pronunciations had given rise to middle and modern ones.

In Gù's time, however, it was not obvious that this was the problem. That he came to see the issue of the contradictions in the reconstructions in terms of the stages of the historical derivation of new from old forms was a breakthrough. The establishment of a historical programme required rejecting what had become a common practice. This was the use of *xiéyin* or *xiéyùn*,[109] 'harmonizing the reading pronunciation' or the 'rhyme category' by transferring the rhyme-words used in a poem, at least semi-arbitrarily, to a category that rhymed in more recent pronunciation, and altering their *fânqiè* 'spelling'.[110] This last was a method, prevalent from about the sixth century CE, of matching the initial and final of the syllable to be 'spelt' with the initial and final of two other syllables whose values were known to the reader. This was not in itself objectionable so long as the derivational trail was preserved, but this was not what had happened. Gù quotes Chén Dì in this regard, in a passage where the historical derivation of pronunciations is compared both to genealogical descent and tracing the course of a river:

> The three hundred pieces [in the *Songs*] are the ancestors of ... reading pronunciations [*yin*]. Those who compose books on the rhymes ought to begin by considering them and trace [the middle

Chinese forms, or the present-day pronunciations] back to their source, and then follow the flow [down from that point] when assigning their written characters to categories. ...

From the Zhou [late second millennium BCE to the third century BCE] some pronunciations had already changed, though these were not the majority. When I investigated the *Shuowén jiêzì* [dictionary by Xû Shèn of the Later Hàn, d. *circa* 121 CE], [I found that] most of them matched the Máo edition of the *Songs*. When, however, Xú Xuân revised the *Shuowén* dictionary, he generally relied on Sun Miân's [*Tángyùn*, a later recension of the] *Qièyùn* [Rhymes by initial and final bisection, a work reflecting the language around 600 CE]. This was using the Táng dynasty pronunciations and violating those of antiquity. Citations from the *Songs* in all the rhyme-books since this date are as rare as morning stars, but there are countless quotations from celebrated authors of the Táng and Sòng periods.

Is this not making a genealogical record (*pû*) of the sons and grandsons while forgetting the ancestors of the lineage?![111]

What Gù objected to was the way that later scholars like Wú Cáilâo of the Sòng dynasty had 'altered the archaic [ascribed] reading pronunciation to make [the words] fit in with Shên Yue's rhymes.'[112] Shên Yue (Xiuwén), who lived from 441 to 513 CE, was perhaps the second writer[113] to produce a manual that showed the four 'tones' (*sheng*) of middle Chinese: level, rising, departing and entering.[114] Gù deplored this failure to respect the historical integrity of information about the past:

> The division of words according to the four tones took place in the period of the Qí and Liáng dynasties [late fifth to early sixth century CE]. At the height of the Zhou dynasty [around the year 1000 BCE or a little later, when most of the pieces in the *Songs* were produced] how could anyone have know about Shên Yue's rhyme-scheme?! I am distressed at how vulgar Confucian scholars of the present time put their trust in the present and have doubts about archaic times.[115]

This sense of history was the basis of one of Gù Yánwû's methodological insights: for the reconstruction of the archaic pronunciations, the phonetic elements of individual characters in a *xiésheng* harmonic series 'cannot be changed,' and the value – whatever it was hypothesized to be – had to be maintained consistently (or within narrow limits) for the characters that used it.[116] Similarly, the principle that had to be adopted when opposing the generally ahistorical nature of the *xiéyin* technique (which was applicable only in a handful of cases) was that 'when a rhyme changed slowly, the men of ancient times did not trouble themselves to alter the written characters'. He approved of the observation made by Chén Zhènsun in Sòng times that 'the archaic and modern ages are different. ... The spoken languages and the tones and pronunciations [*sheng yin*] are in some respects incapable of entirely corresponding the one to the other.'[117]

The first technique that appeared for conveying the approximate pronunciation of a character was *dúruò*, or 'reading as'. Gù observed of this method that

> In Hàn times people did not yet know the method of *fǎnqiè* ... I would further observe that in those cases where the 'is read as ...' given in the original text [that is, the *Shuōwén* dictionary] is not in agreement with the reading [in terms of *fǎnqiè* 'spelling'] of Mr. Xú and of the *Qièyùn* , the pronunciations of the Hàn and Táng had to some extent [already] diverged. If one wishes to seek out the different readings [given to a particular written character] in archaic times and the present day, one uses the Hàn [version] to correct the Táng [version], and the Táng version to correct that of the present day. One relies entirely on the differences between them. If these are not available, one looks to the forms in local dialects [*fāngyán*] for reconstructing the changes, but the ultimate origin [*dǐjí*, literally 'lowest extreme'] is unknowable.[118]

He thus also had a sense of the importance of derivation.

Gù's motivation was not unlike that of some Western seventeenth-century scholars, namely to search for the *prisca*, the lost pristine knowledge once known to the earliest ages. He had a vision of a stable, widespread, and uniform pronunciation of the words in the *Songs* that had endured 'for a period of a thousand and some hundreds of years' from the Sage-Emperor Shùn till the time of the Duke of Zhou.[119] He quotes a work called *Dú Shī zhuóyán* [Unskilled words on reading the *Songs*] that says: 'Theorists maintain that after the Five [races of] Barbarians threw China into disorder [in the fourth century CE] and drove the people of the central plain to the eastern lower Yángzi, and to the northern and southern parts of the region between the Yellow and the Huái rivers, [the language] became mixed with barbarian tongues.' He conceded that 'the changes in sounds and pronunciations [*sheng yin*] may perhaps have begun with this,' but seems to have had the view that following the end of the golden age changes in language were, or had become, an inherent process, since 'even within a single commandery there will be differences in sounds that are related to the location, and within a hundred years speech will have altered in a way that is related to the passing of time.'[120] In his view the loss of the original stable uniformity had started earlier:

> After the Wèi and Jìn dynasties [third to fifth centuries CE] archaic times were remote, and compositions [*cí*] and rhapsodies [*fù*] growing complex. The name later given [to the pronunciation of the finals] was 'rhyme' [*yùn*]. ... But in the literature of the Qín and Hàn [third century BCE to third century CE] the pronunciations had already gradually fallen into disaccord with those of archaic times. With the move to the Eastern Capital [Luòyáng] this became more strongly marked. When [Shên] Xiūwén made his manual of rhymes ... he relied only on the pronunciations used in the rhapsodies of Ban [Gù] and Zhang [Héng], and later writers, and

on those in the lyric verses of Cao [Zhí] and Liú [Zhen] and later writers, selecting from these to make a definitive foundation. Thus the pronunciations of the present day became current, and the ancient pronunciations disappeared. This was a point of change from the view of the study of pronunciations.[121]

This transition from Old to Middle Chinese was followed by a second to what is now called 'Late Middle Chinese':

> Coming down to the Táng dynasty [seventh century to the early tenth], [the state] used [tests in] lyric poetry and rhapsodies to select scholars [for the bureaucracy]. For their books they unvaryingly used the *Qieyùn* of Lù Fâyán [and his colleagues] as their standard.[122] Although this had notes as to which [rhyme-categories] had an independent status [*dúyòng*] and which were used as component parts of composite categories [*tóngyòng*], he did not alter his division into categories [from that of earlier times]. By the Jîngyòu reign-period of the Sòng [1034–7] there was some desire to put this matter in order, but only in the last years of the Emperor Lîzong [1224–64] did Liú Yuan of Píngshûi amalgamate the 206 [archaic] rhymes into 107. [In other words, the phonetic impoverishment of the language since earlier times was at last formally recognized.] The *Yùnhùi* [Compendium of rhymes] by Huáng Gongshào of the Yuán dynasty followed this, and the situation has continued to the present day. The Sòng rhymes have become current and the Táng rhymes have vanished. ... In terms of periods of time, this may be described as 'distance', but in terms of the transmission [of the texts] it must be labelled 'falsification' [*é*]. This was the loss of the [true] Way.[123]

Having the proper pronunciation was essential for understanding the scriptures. He concluded, alluding in passing to a passage in the *Analects* that describes how Confucius was put in danger by the people of the state of Kuang, that: '[Since] Heaven has not destroyed this culture [of ours], it will be necessary for sages to arise once more, and return the pronunciation of the present day to its pure archaic form.'[124] Phonological science was driven by a moral imperative.

The research was based on the analysis of an immense collection of passages from archaic and medieval literature relevant to the pronunciations of particular written characters. Assembling this anthology was a long-term collective creation, with some analogies to a scientific database. The operation took on the nature of a scientific research 'programme' involving many scholars who shared broadly similar objectives. Once the strict adherence to the historic pattern of rhyme-categories, readings 'as', and initial/final bisections that Gù insisted on had been adopted, the challenge was to find a consistent solution to all the equivalences provided by the data. This had a distant resemblance to solving a massive matrix of simultaneous equations.

Consider the entry on *cóng* 'topknot, mane, bristle' in Gù's *Tángyin zhèng* (1:27ab), as a sample of the components constructed for this enterprise. *Cóng* is not in Karlgren's *Grammata Serica Recensa* but was given by Gù as having the 'archaic reading' (*gǔ yin*) of the character that is today read as *cóng* 'a jade piece with a hole in the centre' (Karlgren 1003.g, **dz'óng*, †*dz'uong*). What is interesting is that the initial/final bisectional 'spelling' is given – for Táng times – as what we would today read as *shì* + *jiang* ('warrior-scholar' + '[the Yángzi] river'). In other words, it would yield in modern terms something like *xiang*, not †*dz'uong*. The reconstructed earlier values for these two syllables are **dz'ieg*, †dz'i and **kung*, †kång (where the ' ' ' is a rough breathing and the 'e' is Karlgren's inverted 'e' or shwa). These give a closer, though not perfect, match with the archaic and Táng dynasty values for 1003.g. The focus of Gù Yánwǔ's commentary is on how the 'o' and 'u' vowels in some words, on the one hand, had diverged from the long 'a' vowels (as in *jiang*) in others, and on the other, how in Mandarin (though not so strikingly in Cantonese) they had both had the same rhyme value.

He states that in 'present times,' presumably meaning the Táng dynasty, since this is the focus of the work, that *cóng* and its homophone 'are listed in *both* the categories "second *dong* «winter»" and "fourth *jiang* «river»",' which we can see from the table at the back of his book was a first- or level-tone group. He goes on:

> *Note:* the rhymes *jiang* 'river' [**kung*, †*kång*], *dong* 'east' [**tung*, †*tung*], *dong* 'winter' [**tóng*, †*tuong*] and *zhong* 'wine-vessel' [**tyung*, †*tsyшong*] were [originally] used together. This was still the case under the Northern and Southern Dynasties [fourth to sixth centuries CE]. From Táng times on they began to be inserted in a miscellaneous way under the rhyme *yáng* 'bright force' [**dyang*, †*yang*].
>
> In Sòng times Wú Yù continued this. There was the theory that they [*zong* and its homophone] should both be regularly assigned to the rhyme *yáng* 'bright force'. Under the Yuán dynasty Zhou Déqing, in his *Zhongyuán yinyùn* [pronunciations and rhymes of the north China plain] next had *jiang* 'river' and *yáng* 'bright force' share a *joint* rhyme-category. The *Hóngwǔ zhèngyùn* [corrected rhymes of the Hóngwǔ reign-period [later fourteenth century]] subsequently incorporated *jiang* 'river' into the rhyme-category *yáng* 'bright force'.

We have to remember, when reading this, that Gù (so far as we know[125]) was working without the modern phonetic transcriptions that help us to find our way around the labyrinth and confer important added precision. The use in the Yuán period of the Tibetan hP^h ags-pa alphabet to write the *Méng gǔzì yùn* [Mongol [dynasty] rhymes for the archaic characters][126] seems to have been forgotten.

He ends by quoting earlier scholars to make various points, with chapter and verse cited. The first is that *jiang* 'river' was originally phonetically cognate with *gong* 'work' [**kung*]. The second is that in preimperial works and those from the earliest imperial dynasties where there were rhyme-words used that fell in the *yáng* 'bright force' category, those characters that fell under *jiang* 'river' did *not* appear as rhyming with them, with only a handful of exceptions. Third, *per contra*, already in Hàn times writers were mixing rhyme-words that fell under *dong* 'east' and *dong* 'winter' with words that fell under the categories *yáng* 'bright force' and *táng* 'dynasty'. But the characters containing the character *gong* 'work', such as *gong* 'achievement', could even so be found rhyming with words in the *yáng* 'bright force' category, for instance in the *Lǎozǐ* (a late pre-imperial compilation of possibly earlier materials).

In sum, Gù did not solve the problem, which remains mind-bendingly complex, but he had a clear conception of what the problem was, and of the necessity for a historical-derivational approach to unravel it. And he used it.

THE MATHEMATIZATION OF NATURE

It has been suggested that the essence of the modern scientific revolution in the West was the 'mathematization' of the natural world.[127] This view is not without cogency, but a consideration of the Chinese case indicates the need for a sharper formulation. Representing a selected part of the natural world by a complex of interrelated numerical measurements was not alien to the medieval Chinese way of thinking. Shên Gua's efforts at numerical cartography provide an example from the eleventh century:

> The men of times past had among their books on geography a *Flying Bird Map*, but it is not known who made it. What is meant by 'flying bird' may be explained as follows:
> Even if one has the numbers of *lǐ* for [distances in] the four cardinal directions, they will all have been measured from the number of double paces along the roads.[128] Roads wind or run straight in an inconsistent fashion. If one arranges these distances into a map, the *lǐ* and the double pace measurements will not be consistent with each other. Therefore when one is drawing up a map, one uses a different measurement technique for the *direct* distance in each of the four cardinal directions [which function as x, y coordinates]. This is like a bird flying in a straight line through the air, without any of the distortions caused by going around or adapting one's path to mountains and rivers.
> I have made a *Map for Prefects and County Magistrates*. The scale is 0.2 feet to 180,000 feet [or close to one to a million]. I used the seven methods of the sighting-tube based on a water-level [for triangulation] and cogtooth matching [to determine angles, perhaps especially elevations], the examination of lateral and vertical distances, right angles and acute angles [trigonometry], and curved and straight lines, in order to determine the numbers for [the distances for] a bird in flight. When the map was finished one had the reality [*shí*] of the angles relative to the cardinal directions and the distances [between points], which thus allows the use of the following technique:
> Subdivide the four cardinal directions – which are 8, if the intermediate orientations are included – so as to create 24 directions [which thus requires trisection of angles]. Label them with 8

elements from the Heavenly Trunks series, *jiǎ, yi, bíng, ding, geng, xin, rén*, and *gui*, and with 4 elements from the 64 hexagrams, *qian, kun, gèn*, and *sùn* [12 being enough to define the inclination of the line between two given points]. By these means, even if the map is lost, provided they have the text, later generations can reconstitute it according to these 24 directions and so lay out the configuration of the prefectures and counties.[129]

The amount of surveying needed to make a map at this scale covering a useful area provokes scepticism about whether Shên actually had this work done. (A map half a metre square would cover 250,000 km^2.) It is more likely that he tested the methods but took his data from existing maps, many of which were probably based on rectangular grids by this time.[130] What is interesting is that, for him, the *numbers* were the 'reality'.

The mathematization here is primitive. It refers to a *static* entity, and the numbers are thus *fixed* descriptors. A sharper formulation of what is needed might thus be the 'mathematical functionalization' of the natural world in the sense that a *function* determines how a change in one quantity implies a corresponding change in another quantity according to a specified formula. (Thus the distance (d) travelled in a given time (t) by a uniformly accelerating (a) body starting from standstill is $d = 0.5\ at^2$.)

The Chinese case can still encompass this improvement. Some time before 1584, Zhu Zàiyù, a prince of the Míng dynasty living in self-imposed retirement, discovered the formulae allowing one to tune stringed instruments and pitch-pipes in a well-tempered scale of twelve equal-ratio-interval semitones per octave. The basic story is told in Needham, Wang and Robinson. I recapitulate it here with additional material and emendations to the translations and presentation.[131] The subtleties of a full comparison with the European case are too complex to explore here.[132]

Zhu's work was motivated by a passion for metaphysical metrology that aimed to establish the 'correct' dimensions for the fundamental pitch-pipe, known as the 'Yellow Bell' (*huángzhong*). The pitch-pipe had been conceived of since archaic times as a sort of aether-detector. 'Pitch-pipes [he wrote] are in communication with the aethers [*qì*] of Heaven and Earth, and free of any obstruction; hence the correct pitch emerges from them.'[133] He quoted with approval the view of that 'of all devices only the pitch-pipes are in contact with the Transforming Force [*zàohuà*].'[134] If other measures used to make things, such as those of length, capacity and weight, were calibrated by the correct Yellow Bell, this would make possible a general co-resonating harmony both between humans, and between humans and superhuman beings.[135]

He also argued that in archaic times musicians had had the ability to move without hindrance from a mode based on one fundamental note to another based on another, a capacity later lost. He supported this view by arguing that the obscure term '*caomàn*' in the scriptural *Record of the Rites* (*Lǐjì*) meant 'execute modulations' or 'transpositions', taking the traditional gloss '*zálòng*', literally 'perform music of mixed types' as justifying his interpretation.[136] There is an odour of pious sleight-of-hand in this procedure.

What gave his work a scientific quality, in spite of its archaic metaphysical underpinning, was his insistence on respecting what he called 'the pattern-principles of nature [*zìrán zhi lǐ*]'. These, he said, 'resemble the primary trigrams [of *The Book of Changes*] and are not dependent on the idiosyncrasy of artificial philosophic cleverness.'[137] The foundation was *empirical observation*. In ancient times 'the calculators followed the pitch-pipes in order to determine the algorithm to use, and it was not the case that the pipes were determined by the calculations.' He compared this to a field that grew its grain oblivious of the classification of its shape by the surveyor. Thus 'those who made bells and pitch-pipes in ancient times used their *ears* to mutually calibrate their pitches.' 'People of later times could not do this, and began to rely on numbers to determine their correct proportions.'[138] Such numerology led to its practitioners 'being deluded by numbers and paying no attention to the sounds.' Sounds, he insisted, 'have something that comes from outside measurements and numbers'.[139] In his own work 'the pitch-pipes calculated by the new method are all based on the pattern-principles of nature, after which the numbers are used to seek for a congruence with the sounds. It does not fudge the sounds [it prescribes] to fit with the numbers.'[140] Since equal-temperament tuning is in one sense indeed 'fudging' with respect to intervals based on natural harmonics, in order to create a system in which intervals are consistent with each other, this can sound disingenuous. The answer is to take Zhu's concern for overall consistency as primary.

Zhu tested his results by *experiment* (*shìyàn*).[141] For example, he had pitch-pipes made to the proportions specified by earlier writers and listened to the resulting sounds, which he judged not fully satisfactory.[142] The earlier algorithms had been, he decided, approximations.[143] One reason that people had clung to them, he guessed, might have been comfort with their lack of irrational numbers (*bújìn zhi shù* – 'numbers that never end'). He noted that in his own results (which did contain irrationals) he did not list extraordinarily small quantities beyond a certain point.[144] Nonetheless, *precision* was crucial. Since the craftsmen employed to make bronze pitch-pipes were 'imitating the Transforming Force [*gong móu Zàohuà*],' they had to be highly skilled and carefully supervised, 'paying attention to thousandths and ten-thousandths'.[145] He therefore gave rules and diagrams to guide the making of the twelve-stringed tuning-zither that he had invented.[146]

He noted that according to the twelfth-century philosopher Zhu Xi the practice of players of stringed instruments, such as the seven-stringed zither (*qín*), was not to use any of the old theories but 'the method of the four types of folding and taking the midpoints [*yǐ sìzhé qūzhong zhi fǎ*]'. This involved cutting strips of paper to the length of the strings between their two end-bridges and then folding them to find the points at which harmonic stops had to be marked on the side of the soundboard as a guide to the performer. (Thus the 5/6 and 2/3 points would give the minor third and perfect fifth in natural tuning.) Zhu Zàiyù concurred with Zhu Xi's judgement that 'although this method is in correspondence with the sounds of the pitch-pipes for these sounds, besides being simple and easy to grasp, it is, as regards the procedures and patterns of nature, unclear in that one does not know where [these effects] come from, and is thus, I fear, unavoidably incomplete.'[147]

The problem to be solved is this: ascending and descending sequences of intervals of a fifth that is 'perfect' in the sense of being the exact harmonic of a fundamental frequency create conflicts with the sequences produced by octave intervals. Contrary to the appearance given by the piano keyboard that $7 \times 12 = 12 \times 7$ in semitones, the seventh C produced by octave intervals above a fundamental C is about a quarter of a semitone *lower* than the B# produced by 12 intervals of a perfect fifth starting from the same point. Similar problems occur with fitting in thirds exactly. In Europe two types of compromise tunings ('mean-tone'[148] and an approximate 'equal temperament') were used before the theoretical solution was discovered in the seventeenth century,[149] or, conceivably, borrowed from China. These allowed a fairly smooth modulation from one key to another in, respectively, some and all cases.[150] This practical aesthetic problem does *not* seem to have been a concern among Chinese musicians, who used modes rather than keys and seem to have shifted fundamental notes *between* rather than *in the course of* pieces. The distinction between consonance and dissonance likewise seems not to have been a concern for them. If this impression is sustained by further enquiry, it will be an example of the progress of science being constrained in a quite specific way by cultural factors. Floris Cohen suggests that the shift from Renaissance to Baroque music in Europe *per contra* created strong pressures for the perfection of equal temperament.[151] Zhu's theoretical result was not lost, but had no influence on Chinese musical art.

Zhu was concerned with *transpositions*. One of his inventions was a device to let musicians transpose at sight when using the traditional Chinese heptatonic scale. (This is mappable onto the white keys of a piano as f, g, a, b, c, d, e, f´.) The device was a circular plate around whose circumference ran a band divided into 12 equal segments each labelled with the name of one of the 12 semitones of the octave, starting with *huángzhong*. A smaller disc, mounted on a central spindle, could be rotated by hand inside this band. It was also divided into 12 segments like the slices of a pie. Seven were labelled in consecutive order with the do-re-mi-like names of the seven notes of the heptatonic scale (*gong, shang, jiáo, biànzhí, zhí,*[152] *yú, biàngong*) and the remaining 5 were empty. By turning the central disk to align *gong* with a different member of the 12-semitone sequence, the locations of the other notes could be read off.[153]

The old method for calculating the lengths of the strings needed to generate the 12 semitones of an octave was called *sanfen sûnyì* ('dividing by thirds reduced and augmented').[154] Given an initial length x_1, the sequence was produced term by term by multiplying this length by 2/3 for the first term, then for the second multiplying the preceding result by 4/3, then for the third by 2/3 again, continuing with this alternation, except for x_7 and x_8 for both of which 4/3 was used. To make sense of this, it needs to be borne in mind that the ratio of the relative *frequencies* of the two notes forming a given interval is the *reciprocal* of the ratio defining the *lengths* of their strings. Thus, since 2:3 is the frequency ratio for the natural fifth, the string lengths are in the ratio 3:2. Table 4 lists the values from the old method together with the corresponding values for Zhu's equal temperament, and the ratios for selected intervals of the natural scale.

Table 4   Sanfen sûnyì *and equal-temperament values for the lengths of the resonating strings needed to produce them over one octave as proportions of* x_1

Key on piano	Stage generated by SFSY	SFSY value	Equal temperament value	Natural value ratio	
C	x_1	1.0000	1.0000	1:1	'unison'
C#	x_8	0.9364	0.9439		
D	x_3	0.8889	0.8909	8:9	'tone'
D#	x_{10}	0.8324	0.8409	5:6	'minor 3rd'
E	x_5	0.7901	0.7937	4:5	'major 3rd'
F	x_{12}	0.7399	0.7492	3:4	'fourth'
F#	x_7	0.7023	0.7071		
G	x_2	0.6667	0.6674	2:3	'fifth'
G#	x_9	0.6243	0.6300	5:8	'minor 6th'
A	x_4	0.5926	0.5946	3:5	'major 6th'
A#	x_{11}	0.5549	0.5612		
B	x_6	0.5333	0.5297	8:15	'seventh'
Ć	x_{13}	0.4933	0.5000	1:2	'octave'

The *sanfen sûnyì*, in spite of weaknesses like its imperfect octave – which practitioners corrected in practice – was already a non-trivial mathematization of the natural world that probably dates to the preimperial age.[155] More than a millennium-and-a-half earlier, the historian Sima Qian had declared that 'when the numbers take on form they become musical sounds'.[156] The general attitude was thus an old one, though Zhu himself took the more abstract view that 'numbers are entities that have no forms'.[157] This sums up one of the distinctions between a numerology and a mathematization.

For his equal-temperament scale Zhu used what is still the standard method, dividing the length of the string or pipe giving the fundamental note ('yellow bell') successively by the twelfth root of two ($2\char`^(1/12) = 1.059463\ldots$). For the pipes he further compensated for the end-effect[158] by dividing each diameter successively by the twenty-fourth root of 2 ($= 1.029302\ldots$). He summed up what he had done as follows:

> Pitch is generated by numbers. If the numbers are correct, the pitches will all be in accord with each other. If it should chance that a pitch is not in accord [with the others] its number is not correct. Someone who has reached the pattern-principle of the numbers behind musical pitches has gained his intellectual access *by adapting to how they change*, it not being possible to hold on to a single

[preconceived] idea. For this reason I do not use the method of dividing by thirds with reduction and augmentation [*sanfen sûnyì zhi fã*]. I have founded a new procedure. I establish the length of one foot as the dividend (*shí*), and use the secret ratio (*mìlü*) to divide it. *In all, this is done 12 times.* The true numbers for the [lengths of the] pitch-pipes that I have sought are much simpler and swifter than the four old types of method.[159] The numbers and the sounds on the zither mutually confirm each other, tallying to an extreme degree.[160]

There is of course only one operation that applied twelve times gives the desired result.

Although he called the twelfth root of two his 'secret ratio', Zhu shows how its value can be derived using only the extraction of square roots and cube roots. The account he gives in his other major work *The Quintessential Significance of the Standard Pitchpipes* is made unnecessarily obscure by his pretence that what he is doing is the recovery of ancient wisdom from the *Ritual of Zhou* (*Zhoulî*).[161] Stripped of its ideological fancy-dress all he does is to calculate as follows: consider 2 octaves, an upper and a lower, with a middle *huángzhong*, or fundamental, having a length of 1 foot of 10 inches; find the square root of the *huángzhong* one octave below it, that is, of length 2; this is 1.4142,1356. ... to 24 places of decimals; this is also the ratio giving the length of the midpoint note, *rúibin*, in the lower octave and twice the length of the *rúibin* in the upper octave (0.7071678 ... feet as in Table 4). Take the square root again, which gives 1.189207 ... feet, which is the length for the note *nánlü* in the lower octave, and half of that for *nanlü* in the upper octave (0.5946036 ...). Take the cube root of the result, which gives 1.059463094 .... This is the length of the *yìngzhong*, or penultimate note, in the lower octave. This is the 'secret ratio', the twelfth root of 2 (since $2^{(1/12)} = 2^{(1/2)^{(1/2)^{(1/3)}}}$), and twice the length of the *yìngzhong* an octave above. Dividing this length by the twelfth root of two, that is, *by itself*, evidently yields 1, the length of the middle *huángzhong* or fundamental. From this he derived the procedure needed to move up the equal-tempered scale a semitone at a time. He concludes: 'For this reason, for each pitch one takes it as the true measure of the *huángzhong* [for the purposes of this calculation], and multiplies it by 10 inches to be the dividend, and then divides it by the value of the *yìngzhong* of the lower octave [i.e. the twelfth root of 2] to obtain the next pitch upwards [in inches].' This was an 'endless cycle'. His shift to using inches as the units was also unnecessary.[162]

The above is a premodern Chinese case of the 'mathematical functionalization' of the natural world.

THE SOCIAL MATRIX

What happened to Zhu's equal-temperament system once discovered? He wrote tunes for what he called the 'style of shifting the fundamental' (*xuángong fã*) that it made possible. In other words a sort of 'modal modulation'.[163] His books and tablatures, for the most part written between 1567 and 1581, were printed in 1595.[164] His ideas and melodies thus survived. In the mid-Qing, his system was the subject of ferocious attack in a musicological compendium issued under imperial sponsorship. He had, however, a heavyweight defender in Jiang Yǒng (d. 1762) who stated, regardless of the aura of imperial authority surrounding the opposite view, that, 'Most of those who came after Zhu Zàiyù did not grasp his meaning or made reckless criticisms of minor defects.'[165] Since his work in a way completed a 'programme' that had been in existence for slightly over a millennium, namely to find a system of equal temperament rather than just *ad hoc* adjustments,[166] it is surprising that his achievement was not immediately acclaimed. Looking briefly at this issue casts light on how far late-imperial China was, or was not, a society favourable to scientific endeavour.

When Zhu submitted his works to the Wànlì emperor in 1595 he was aware that what he was propounding was explosive. He wrote in his accompanying memorial:

> The study of tuning has been perverted for a long time. This is because people have held in honour as models the absurdities of the three theories of a Yellow Bell of 9 inches [rather than a foot of 10 inches], the method of tuning by 'dividing by thirds reduced and augmented', and the cycle of fifths [*géba xiang sheng*].[167] Although these three theories are indeed absurdities in acoustical science, the entire world holds them in honour as models. It is also to be explained by the fact that there have only been a few people who have heard your minister's [my] ideas and not thought him to be preposterous. This is why your minister, in his simpleminded way, for all that he has possessed them in his own mind, has kept his lips sealed and for many years held them secret, not daring to give them written form.[168]

The emperor commented that 'it is evident that the petitioner has given attention to music and tuning, and is highly to be commended. The three works that have been presented are to be retained for examination [*lán*].' One set each was sent to the Ministry of Rituals, the National University,[169] the Hànlín Academy, and the Hall of Literary Profundity (*Wényuangé*), the governmental library. No official appraisal was made of Zhu's books, and it does not seem that one was explicitly ordered: besides 'examine', *lán* only means to 'look at', 'read' or 'inspect'. They would have threatened the prestige of Court musicologists and musical professionals, and this may partly explain the ensuing silence. Zhu insisted that his work 'differs to a considerable extent from the older theories of former Confucian scholars,' cited with approval his father's dictum that 'through successive dynasties those who have created music have been unable to do it properly,' and pointed to the long-standing loss of a tradition that he saw as having been vital for maintaining people's physical and emotional health as well as rapport with the spiritual world.[170] For a professional in the Court of Imperial Sacrifices or the Office of Music, charged with putting on performances for ritual and ceremonial occasions, to have accepted his point of view would have been accepting an indictment of incompetence of a nature that could lead to politically dangerous consequences for the dynasty.

The first recorded attack only came a century and a half later, under the Qing dynasty, in the *Second Compilation on the Basic Principles of the Tuning Pipes* (*Lülü zhèngyì hòubian*), compiled at the order of the Qianlóng emperor and printed in 1746. The chief editor, Zhang Zhào, was a calligrapher, poet and dramatic musical composer, of some real ability.[171] He and his colleagues noted Zhu's 'skilfulness in calculating' and showed that they understood what he had been doing when they said he had been 'making mischief [nì]' by using twice the value of the Responding Bell [yìngzhong, that is, the twelfth root of two] to divide each note so as to generate the equal-tempered sequence. The series should have been derived from the Yellow Bell, which was primary and so gave the standard pitch. In all he was guilty of ten 'great crimes [zùizhuàng]' in putting forward this 'notion without foundation [yìshuo]'. He had 'abandoned the Confucians of ancient times', 'trusted to his own wisdom and created a new theory', 'maintained that the pitches were determined by the sounds and not by the measures', and held views contrary to various ancient quasi-scriptural authorities. There was not one iota of scientific refutation, but they were right in exposing his attempt to dress up his system as a return to the practice of antiquity. Later, the compilers of the summaries made of over ten thousand books as part of the preparations for the project that resulted in the *Complete Works of the Four Treasuries* (*Sìkù quánshu*) in 1781 picked on this in order to dismiss his ideas (which they probably did not understand) with contempt.[172]

Zhu Zàiyù was at least to some extent driven by ideological motives, as emerges from his comment in *A New Theory of Musicology* (*Yuèxué xin shuo*) that 'if we use the method of shifting the fundamental [modal modulation] to compose new scores in a manner resembling [the past] [nìzào xin pú] this will enable those who study musical pitches in generations to come to see them and become deeply convinced that the music of antiquity is still manifestly in existence, and that its transmission has not been lost.'[173] At the end of some tablatures he had made of contemporary popular music he observed that

> Musical notes [yin] arise because they are produced in the heart-minds of human beings. The heart-minds of human beings are no different today from what they were in antiquity, so how could the musical notes differ? If we set the phrases of antiquity [like those in the *Scripture of Songs*] to the tunes of the present day this will make them easy for people to understand and to become aware of the pattern-principles in the music. How admirable![174]

His science was motivated, as well as obscured, by a Confucian zeal that contained elements both of reform and renaissance. The implication was that the authorities had got it wrong.

While Zhu's theories still commanded some attention in the eighteenth century, they were not widespread. The phonologist, mathematician and musicologist Jiang Yông had to search for a long time to locate Zhu's works and only obtained a set in the closing years of his life. On reading them, he was, he says, 'both startled with horror and so happy I leapt!' He was convinced immediately and noted that Zhu's values, when compared to those derived by the old method of 'dividing by thirds reduced and augmented', 'only differ in the second and third places of decimals'. Precision mattered. He then wrote a book filling in gaps in Zhu's system and making minor improvements. So far as he was concerned it was unambiguously a 'new method'.[175]

Both Zhu and Galileo in their different ways challenged cosmological orthodoxy. Both had their science right. Zhu's results may have been the more accurate. Both were accused of crimes by the ideological authorities, though Zhu only posthumously. Galileo's views triumphed. Zhu's faded away, though without violent repression. The differences in the European and Chinese social matrices that helped or hindered science need subtle pondering.

Acknowledgements

Apart from noting my fundamental debt to the late Alistair Crombie, I should also like to thank Floris Cohen and Toby Huff for their comments on the revised draft of this essay. They take different positions from mine on most of the issues discussed, but this makes their criticisms more rather than less valuable. I am also indebted to Kenneth Robinson for his helpful observations on the section on acoustics.

Notes and References

1. A. Crombie, *Styles of Scientific Thinking in the European Tradition. The history of argument and explanation especially in the mathematical and biomedical sciences and arts* (London, 1994), 3 vols.

2. Crombie, *Styles* (1), 46.

3. R.M. Grove, *Green Imperialism: Colonial Expansion, Tropical Island Edens and the Origins of Environmentalism, 1600–1860* (New York, 1994), esp. 355, 360–4.

4. R. Iliffe, 'Rational Artistry', *History of Science* 36, 1998.

5. M. Elvin, 'Personal Luck: why premodern China – probably – did not develop probabilistic thinking'. Paper deposited with the Needham Research Institute, 8 Sylvester Road, Cambridge, UK, 2000. Scheduled for publication in 2004 in a volume edited by G. Dux and H.U. Vogel. There is a Chinese version in Liú Dùn and Wáng Yángzong (eds.), *Zhōngguó kèxué yǔ kèxué gémìng. Lî Yuesè nántí jí qí xiāngguan wèntí yánjiù lúnzhù xuân* [Chinese science and scientific revolution. Selected research on Needham's conundrum and related questions] (Shenyáng, 2002).

6. Tones are shown for modern characters as follows: 'a' – level (unmarked), also used for neutral tones; 'á' – rising; 'ǎ' – dipping-rising; 'à' – falling and dynamically *fp*. Where the tone in some personal and place names can be read in one of two ways, I have not always been able to determine which is correct. Tones for premodern words are not shown.

7. For the main participants see references to Chén Dì, Gù Yánwǔ*, Dài Zhèn*, Qián Dàxin*, Duan Yùcái*, Wáng Niànsun*, Jiang Yôugào, Yán Ruòju*, and Cúi Shù* in A. Hummel (ed.), *Eminent Chinese of the Ch'ing Period (1644–1912)* (Washington, 1943–4), 2 vols. (An asterisk indicates an entry dedicated to the scholar concerned.) There is a survey in B.A. Elman, *From Philosophy to Philology: Intellectual and Social Aspects of Change in Late Imperial China* (Cambridge, Massachusetts, 1984), 212–21. He sees the historicization of approach as being particularly the achievement of Jiang Yông (1681–1762).

8. M. Elvin, 'Who was responsible for the weather? Moral meteorology in late imperial China', in M.F. Low (ed.), *Beyond Joseph Needham, Technology, and Medicine in East and Southeast Asia, Osiris* 13, 1998; and M. Elvin, 'The Man Who Saw Dragons: Science and Styles of Thinking in Xie Zhaozhe's *Fivefold Miscellany*', *Journal of the Oriental Society of Australia*, 25 and 26, 1993–4.

9. Elvin, 'Personal luck' (5).

10. Zhu Zàiyù [Míng], *Lüxué xinshuo* [A new theory of the science of the pitch-pipes], Féng Wéncí (ed.) (annotated reprint of the 1584 edition: Bêijing, 1986), 19.

11. M. Elvin, 'The unavoidable environment: reflections on premodern economic growth in China,' in Ts'ui-jung Liu [Liú Cùiróng] and Shou-chien Shih [Shí Shôuqian] (eds.), *Economic History, Urban Culture and Material Culture* (Táibêi, 2002).

12. E.S. Rawski, *Education and Popular Literacy in Ch'ing China* (Ann Arbor, 1979).

13. For example, E. Eisenstein, *The Printing Press as an Agent of Change* (Cambridge, 1980), or A. Johns, *The Nature of the Book: Print and Knowledge in the Making* (Chicago, 1998).

14. M. Elvin, 'Skills and resources in late traditional China', in D. Perkins (ed.), *China's Modern Economy in Historical Perspective* (Stanford, 1975), reprinted in Elvin, *Another History. Essays on China from a European Perspective* (Sydney, 1996).

15. E.g., J. Besson, *Theatrum Instrumetarum et Machinarum* (Paris, 1573); A. Ramelli, *Le diverse et artificose machine del capitano Agostino Ramelli* (Paris, 1588); and G.A. Böckler, *Theatrum machinarum novum* (Nuremberg, 1662).

16. I. Newton, *The Principia. Mathematical Principles of Natural Philosophy*, translated by I. Bernard Cohen and Anne Whitman (Berkeley, 1999), 418.

17. Discussed in footnote 68 of M. Elvin and N. Su, 'Action at Distance. The Influence of the Yellow River on Hangzhou Bay since A.D. 1000', in M. Elvin and T-J. Liu, (eds.), *Sediments of Time. Environment and Society in Chinese History* (New York, 1998).

18. Sòng Yìngxing, 'Lùn qì' [A discussion of matter-energy], in id., *Yêyì, Lùn qì, Tán Tian, Silián shi* [Unofficial policy papers, A discussion of matter-energy, On the heavens, and Poems of pity] (Shanghai, 1976), 68. On Sòng generally, see C. Cullen, 'The Science/technology interface in seventeenth-century China: Song Yingxing on *qi* and the *wuxing*', *Bulletin of the School of Oriental and African Studies* 53.2, 1990.

19. Elvin, 'Skills' (14), and M. Elvin, 'Blood and Statistics: Reconstructing the Population Dynamics of Late Imperial China from the Biographies of Virtuous Women in Local Gazetteers', in H. Zurndorfer (ed.), *Chinese Women in the Imperial Past: New Perspectives* (Leiden, 1999).

20. Wáng Jiànqiu, *Sòngdài Tàixué yû Tàixuésheng* [The Great School and its students in the Sòng dynasty] (Táibêi, 1965). See also J. Needham and Lu Gwei-djen [Lu Gùizhen], 'China and the origin of examinations in medicine', *Proceedings of the Royal Society of Medicine* LVI.2, Feb. 1963.

21. E. Grant, *The Foundations of Modern Science in the Middle Ages. Their Religious, Institutional, and Intellectual Contexts* (New York, 1996).

22. T. Grimm, 'Academies and Urban Systems in Kwangtung', in G.W. Skinner (ed.), *The City in Late Imperial China* (Stanford, 1977).

23. G. Bachelard, *La formation de l'esprit scientifique* (Paris, 3rd ed., 1957).

24. W. Eamon, *Science and the Secrets of Nature. Books of Secrets in Medieval and Early Modern Europe* (Princeton, 1994).

25. On Xiè see Elvin, 'Dragons' (8).

26. Even he got the wrong empirical numerical results as Mersenne complained from Paris. See Crombie, *Styles* (1), 675, 817–19.

27. S. Kuznets, *Modern Economic Growth. Rate, Structure, and Spread* (New Haven, Connecticut, 1966), 9. Kuznets provides useful benchmark tests for 'modern' economic growth, such as the decline of the proportion of the labour force engaged in primary production (essentially agriculture and stockraising).

28. M. Elvin, *The Pattern of the Chinese Past* (Stanford, 1973).

29. W. Van O. Quine, *Set Theory and Its Logic* (Cambridge, Massachusetts, 1963), 2–3 and ch. 11.

30. A.C. Graham, *Later Mohist Logic, Ethics and Science* (London, 1978), 40, 440–1.

31. Graham, *Mohist Logic* (30), 58.

32. Retranslated from Graham, *Mohist Logic* (30), 439. See M. Elvin, 'The Logic of Logic. A Comment on Mr. Makeham's Note', *Papers on Far Eastern History* 42, 1990.

33. C. Harbsmeier, C., *Science and Civilisation in China* VII.1, Language and Logic (Cambridge, 1998), 358–408.

34. K. Chemla, 'Relations between procedure and demonstration. Measuring the circle in the *Nine Chapters on Mathematical Procedures* and their commentary by Liu Hui (3rd century)', in H.N. Janke, N. Knoche and M. Otte (eds.), *History of Mathematics and Education: Ideas and Experiences* (Göttingen, 1996); K. Chemla, 'What is at stake in mathematical proofs from third-century China?', *Science in Context* 10.2, 1997; and K. Chemla, 'Fractions and irrationals between algorithm and proof in ancient China', *Studies in the History of Science and Medicine* XV.1–2 (1997–8).

35. C. Jami, *Les Méthodes Rapides pour la Trigonométrie et le Rapport Précis du Cercle (1774). Tradition chinoise et apport occidental en mathématiques* (Paris, 1990), 41.

36. P. Engelfriet, 'The Chinese Euclid and its European Context', in C. Jami and H. Delahaye (eds.), *L'Europe en Chine. Interactions scientifiques, religieuses et culturelles aux XVIIᵉ et XVIIIᵉ siècles* (Paris, 1993).

37. Xiè Zhàozhè, *Wŭ zázŭ* [Fivefold miscellany] (1608. Reprinted: Táibèi, 1971, 2 vols.), 360-1. There is another account at *ibid.* 272 stressing the 'minute particularity' with which the seafarers saw the dragons.

38. Xiè Zhàozhè, *Miscellany* (37), 127. For background, see J. Needham and Lu Gwei-djen [Lu Gùizhen], 'The earliest snow crystal observations', *Weather* XVI.10, Oct. 1961.

39. What Lî says is 'I have cut open the stomach [whether of one or more is not clear] and there are, in approximate terms, somewhat over a catty [*jin*] of ants [inside].' See Li Shízhen, *Bĕncăo gangmù* [Pharmacopeia arranged by headings and subheadings] (1596, reprinted, Shangwù yinshuguân edition: Shànghâi, 1930, 5 vols. and index), 22:60, *juàn* 43. On another edition of this work see note 68.

40. J. Needham, Lu Gwei-djen [Lu Gùizhen] and Huang Hsing-tsung, *Science and Civilisation in China*, VI.1: Botany (Cambridge, 1986), 318. What Lî says is: 'The old prescriptions say that soya-beans dissolve the effects of numerous drugs and poisons [*bâi yàodú*], but every time I tried this they have been quite ineffective. If one adds "liquorice" [*gancăo*] their efficacy is remarkable. It is essential to know this.' Lî Shízhen, *Pharmacopeia* (39), 16:90, *juàn* 24.

41. Xiè Zhàozhè, *Miscellany* (37), 1080–1.

42. *Shícháo Shèngxùn* [The sacred instructions of ten reigns] (99 vols, 1616 to 1874, with prefaces from 1666 to 1880. Referred to in the text by reign-period. Publication details are not specified, but the work was presumably issued by imperial order in Bĕijing during the Guangxù reign-period), Yongzhèng 8:1ab.

43. *Sacred Instructions* (42), Yongzhèng 8:7a.

44. Yáng Qîqiáo, *Yongzhèng-dì jĭ qí mìzhé zhìdù yánjiù* [The Yongzhèng emperor and his system of secret memorials] (Hong Kong, 1981), 27.

45. *Sacred Instructions* (42), Yongzhèng 8:3b.

46. These issues are discussed in more detail in Elvin, 'Weather' (8, 1998).

47. J. Needham, Ho Ping-yü [Hé Bìngyù] and Lu Gwei-djen [Lu Gùizhen], *Science and Civilisation in China* V.3: Chemistry and Chemical Technology: Part 3: Spagyrical discovery and invention: historical survey, from cinnabar elixirs to synthetic insulin (Cambridge, 1976), 108, make the intriguing off-the-cuff remark about Gé Hóng (3rd to 4th century CE), 'How did he manage to make so many true observations of chemical behaviour, and carry out so many interpretable experiments, even though he himself could never interpret them?' It has to be asked to what extent they involved 'experimental' thinking in the relatively strict sense given to the term in the present discussion. It needs to be distinguished, for example, from the everyday sense of 'experimentation' in the sense of 'trying out' – new recipes, for example.

48. Xiè Zhàozhè, *Miscellany* (37), 940.

49. Xiè Zhàozhè, *Miscellany* (37), 857.

50. Tián Wén, (Qing), *Qián shu* [The book of Gùizhou] in Yan Yiping (ed.), *Yuèyâ-táng cóngshu*, case 36 in *Bâibù cóngshu jíchéng* [Collection of collections on all topics] (reprinted by Yiwén Yinshuguân: Táibĕi, 1965–8, in 830 cases), 4::4:5ab. The '::' in citations indicates the number preceding is that of the 'case'. I have found Lombard-Salmon's translation very useful, even if I have often, with hesitations, differed from it. See C. Lombard-Salmon, *Un exemple d'acculturation chinoise: la province de Guizhou* (Paris, 1972), 190–1.

51. Twenty *sheng*. The Qing *sheng* was 1035 ml.

52. G. Bauer, [Agricola, Georgius], *De Re Metallica* (1556), H.C. and L.H. Hoover, transl. and ed. (1912. Reprinted, New York, 1950), notably the methods on 430–2.

53. I owe these explanations to Dr Ian Williams of the Research School of Earth Sciences, Australian National University, and would like to express my thanks.

54. Xiè Zhàozhè, *Miscellany* (37), 68–9. Xiè also used other arguments: a quotation from scripture, and theoretical considerations based on *yin-yáng* theory.

55. Sòng Yìngxing, 'Lùn qì' (18), 72, 69 and 70 respectively.

56. Personal observations in 1994 of the 70-metre scale model of the inner Hángzhou Bay, Zhèjiang Province, used for simulations in Hángzhou.

57. Here, presumably the seasons when certain plants blossomed or migrant birds appeared.

58. J. Needham and Wáng Líng, *Science and Civilisation in China* III: Mathematics and the sciences of the heavens and the earth (Cambridge, 1959), 359. Notes and transliterated Chinese characters omitted.

59. J. Needham and Wáng Líng, *Science and Civilisation in China* III (58), 361. On the history of the mechanical details see J. Needham and Wáng Líng, *Science and Civilisation in China* IV.2: Physics and physical technology, part 2, Mechanical engineering (Cambridge, 1965), section 27j.

60. Shi Sù *et al.*, (eds.), revised by Shên Zuòbin, *Jiatài Gùiji zhì* [Jiatài reign-period gazetteer for Gùiji] (Reprinted in vol. 7 of *Sòng-Yúan fangzhì cóngkan* [Collection of Sòng- and Yúan-dynasty local gazetteers] (Bêijing, 1992), 7065. See also Needham and Wang, *Science and Civilisation in China* III (58), 491.

61. The term *qi* in this phrase properly means 'stones on the side of a mountain', or scree.

62. Zhang Hào (ed.), *Bâoqìng Gùiji xù-zhì* [Bâoqìng reign-period continuation of the gazetteer for Gùiji] (Reprinted in vol. 7 of *Sòng-Yúan fangzhì cóngkan* [Collection of Sòng- and Yúan-dynasty local gazetteers] (Bêijing, 1992), 7180.

63. J. Brown *et al.*, *Waves, Tides and Shallow-Water Processes* (Oxford, 1989. Produced by the Open University), 62.

64. W. Debler, *Fluid Mechanics Fundamentals* (Englewood Cliffs, 1990), 250–1.

65. Elvin and Su, 'Hangzhou Bay' (17), 368 and 370.

66. Crombie, *Styles* (1), II, ch. 15.

67. Grove, *Green Imperialism* (3); K. Reeds, 'What the Nahua knew. The secrets of Mexico's plants had been revealed by the sixteenth century', *Nature* 6879 (28 Mar. 2002), 369–70.

68. References are to the five-volume Shangwù yinshuguân edition published in Shànghâi in 1930 (as in note 39). Readers may however find Lǐ Shízhen, *Bêncâo gangmù:* [Pharmacopoeia arranged by headings and subheadings] (1596. Reprinted, Rénmín wèisheng chubânshè: Bêijing, 1975. 4 vols). easier to access. *Juàn* numbers have been included in these references to facilitate location of passages.

69. N.R. Scott-Ram, *Transformed Cladistics, Taxonomy and Evolution* (Cambridge, 1990).

70. Crombie, *Styles* (1), II, 1267.

71. W. Blunt, *The Compleat Naturalist. A Life of Linnaeus* (first published in 1971, reprinted, London, 2001), appendix I by W.T. Stearn, 246–8.

72. *Quèxíng* here might also be translated as 'move backwards', but it is hard to see how this could make any zoological sense. *Què* can mean *yâng*, 'face upwards', and this gives a more intelligible rendering.

73. Lǐ Shízhen, *Pharmacopeia* (39), 21/55, *juàn* 39.

74. Lǐ Shízhen, *Pharmacopeia* (39), 6/33, *juàn* 3. I use the translation 'quasi-cholera' as it is unlikely that the disease called *hùoluàn* before the 1820s was true classic cholera. See K. MacPherson, 'Cholera In China, 1820–1930: An Aspect of the Internationalization of Infectious Disease', in Elvin and Liu (eds.), *Sediments of Time. Environment and Society in Chinese History* (New York, 1998). Emphasis added.

75. Lǐ Shízhen, *Pharmacopeia* (39), 21/89–90, *juàn* 40.

76. Lǐ Shízhen, *Pharmacopeia* (39), 6/32–3, *juàn* 3. The last example has been very slightly rearranged so as to show the structure more clearly. Emphases added.

77. Lǐ Shízhen, *Pharmacopeia* (39), 6/34, *juàn* 3.

78. Lǐ Shízhen, *Pharmacopeia* (39), 6/36, *juàn* 3.

79. This provisional conclusion is based mainly on P. Huard and M. Wong, 'Évolution de la matière médicale chinoise', *Janus* 47, 1958, 45–50.

80. The answer is 2: a 5 and a 6, or a 6 and a 5. Leibniz is said to have failed to solve this correctly.

81. I am grateful to Professor Jochi Shigeru for drawing this passage to my attention.

82. Bái Shàngshù, '*Jiŭzhang suàn shù' zhùshì* [Annotated explanations of *Nine Chapters on the Mathematical Art*] (Bêijing, 1983), 21–3; Qián Bâocong (ed.), *Suânjing shíshu* [Ten mathematical classics], II, *Jiŭzhang suànshù* [Nine chapters on the mathematical art], with commentary by Liú Hui (3rd c. CE) (Bêijing, 1963), 71–2; Wú Wénjùn (ed.), *Zhongguó shùxué dàxi* [Compendium on Chinese mathematics] (Bêijing, 1998), Volume 2; Shên Kangshen (ed.), *Zhongguó gûdài shùxué míngzhù 'Jiŭzhang suànshù'* ['Nine Chapters on the Mathematical Art': A masterpiece of ancient Chinese mathematics], II.59–60. On text of the 'Nine Chapters' see J-C. Martzloff, *A History of Chinese Mathematics* (original edn., 1987; rev. English edn., Berlin and Heidelberg, 1997), translated from the French by S.S. Wilson, 127–36.

83. Qián Bâocong, *Nine Chapters* (82), 6:1–2. Bái Shàngshù (82),184–7 gives the method. See also Wú Wénjùn/Shên Kangshen, *Nine Chapters* (82), II.85–6.

84. The exact meaning of the terms in quotes in this sentence is not relevant here, and they should be taken as approximations only.

85. There is a brief history in Wu Kang, *Outline of the Changes of the Zhou*. For the general background, see Needham and Wang, *Science and Civilisation in China* II: History of Scientific Thought (Cambridge, 1956).

86. R. Wilhelm, *The I Ching or Book of Changes*, transl. C.F. Baynes (New York, 1950, 2 vols), I.xl–xli.

87. See Lî Líng, *Zhongguó fangshù kâo* [A study of magical techniques in China] (rev.ed., Bêijing, 2000), ch. 4.

88. Wú Kang, *Shào-zî Yì-xué* [Master Shào's study of the *Changes*] (Táibêi, 1959), 37.

89. Chéng Shìquán, 'Zhou Yì chéng guà jí Chun-qiu shìfâ' [The transformation of the *Changes of Zhou* into hexagram form and methods of divining by plants during the Springs and Autumns period], in Huáng Shòuqí and Zhang Shànwén (eds.), *Zhou Yì yánjiù lùnwénjí* [Collected essays on the *Changes of Zhou*] (Bêijing, 1989), II, 377.

90. Sometimes given as *ptarmica* or *sibirica* by earlier writers.

91. I follow the formulation given by Wilhelm, *The I Ching* (1950), 86.

92. There is a version of this table, with other matter, in Wáng Qí, *San Cái túhùi* [Illustrated encyclopaedia of the Three Principles] (1607. Reprinted Táibêi, 1970, 6 vols.), vol. 4, 10:18ab,/1795–6. The three differing derivations of the young yin and young yáng are each seen as having a distinctive nature. They are each described by one of the hexagrams that is the reduplicated form of a trigram. Thus (9, 4, 8) is symbolized by *kân* 'the abysmal'.

93. Needham and Wang, *Science and Civilisation in China* III (58), 140, hint at the existence of 'an esoteric doctrine', without elaborating further.

94. Pascal's triangle starts with a '1' at its summit, and each of the numbers in each row below is derived as the sum of the two numbers offset half a position to each side of it in the row above: (1), (1, 1), (1, 2, 1), (1, 3, 3, 1), etc. The numbers in each row sum to 2 raised to the power of the rank of the row, starting with 0: thus $\sum(1, 3, 3, 1) = 2^3 = 8$.

95. Martzloff, *Chinese Mathematics* (82), 17, 230-1, 304-5.

96. Luó Xinbên and Xû Róngsheng, *Zhongguó gûdài dûbó xísú* [Gambling customs in ancient China] (Xi'an, 1994), 106–8.

97. Mèng Yuánlâo, *Dongjng mènghuà lù* [Dream of the Glories of the Eastern Capital], with notes by Dèng Zhichéng (Bêijing, 1959), 162. I have also consulted the Japanese translation by Iriya Yoshitake and Umehara Kaoru, *Tôkyô muka roku* (Tokyo, 1996), 199.

98. Note cited by Dèng Zhichéng in Mèng Yuánlâo, *Dream* (97), 164.

99. Note cited by Dèng Zhichéng in Mèng Yuánlâo, *Dream* (97), 163.

100. Note cited by Dèng Zhichéng in Mèng Yuánlâo, *Dream* (97), 163–4.

101. Lî Kuangyì, *Zixià lù* [*Providing For Leisure Time*], in *Qindìng sikù quánshu* (Táiwan shangwù yinshuguân edn.: Táibéi, 1983), *zibu* 10, vol. 850, 157.

102. Guó Shuanglín and Xiào Méihua, *Zhongguó dûbó shî* [A history of gambling in China] (Bêijing, 1995), 181–3.

103. This reconstruction is based on the most coherent compromise I can devise from the somewhat conflicting accounts in Guó and Xiào (104), and in Luó and Xû (98).

104. We have also to consider the possibility that the rules may have the special feature that the principal had to be the first of the pair. In other words: 2′. Permitted: 1**2**, 2**3**, 3**4**, and 4**1**. *Not permitted:* 1**2**, 2**3**, 3**4**, or 4**1**; 1**3**, 1**3**, 2**4**, or 2**4**.

105. Gù Yánwû, *Yînxué wû shu* [Five works on phonology] (seventeenth century. Reprinted, Sixián jiàngshè: n.p., 1890, 12 vols.). There is a modern single-volume reprint in the *Yînyùnxué cóngshu* [Series on phonology] published by the Zhonghuá shujú, Bêijing, in 1982.

106. E. Pulleyblank, *Lexicon of Reconstructed Pronunciation in Early Middle Chinese, Late Middle Chinese, and Early Mandarin* (Vancouver, 1991), divides Middle Chinese into Early and Late.

107. B. Karlgren, *Grammata Serica Recensa* (Stockholm, 1957), 650a and 663a. Karlgren's phonetic typography cannot be reproduced here without fonts not available to me. See also Gù Yánwû, *Pronunciations* (105), zhong:1b. I have used 'ɛ' to represent Karlgren's inverted 'e'.

108. Gù Yánwû, *Pronunciations* (105), zhong:1a. I have not included the more recent reconstructions in Pulleyblank, *Lexicon* (106) as these, though superior, do not affect the argument put forward here.

109. This *xié* is Karlgren, *Grammata Serica* (107), 639c ,and different from the *xie* in *xiésheng* given earlier, which is *ibid.*, 599g.

110. Described by Pulleyblank, *Lexicon* (106), 20, as 'the scholastic invention of special readings of characters to improve the rhyming of the *Shijing* that was practised in the Sòng dynasty.'

111. Gù Yánwû, *Pronunciations* (105), zhong: 6a.

112. Gù Yánwû, *Pronunciations* (105), zhong: 1b–2a.

113. The first was probably Zhou Yong in the 5th century CE.

114. Not of course the same as the four tones of modern standard Chinese.

115. Gù Yánwû, *Pronunciations* (105), zhong: 2a.

116. Gù Yánwû, *Pronunciations* (105), zhong: 3a.

117. Gù Yánwû, *Pronunciations* (105), zhong: 1a.

118. Gù Yánwû, *Pronunciations* (105), xià: 11ab.

119. Gù Yánwû, *Pronunciations* (105), author's preface: 1ab.

120. Gù Yánwû, *Pronunciations* (105), zhong: 6a.

121. Gù Yánwû, *Pronunciations* (105), author's preface: 1b.

122. A feature of the *Qièyùn* was that it differentiated words according to their tone.

123. Gù Yánwû, *Pronunciations* (105), author's preface: 1b–2a.

124. Gù Yánwû, *Pronunciations* (105), author's preface: 2b.

125. It has been suggested by Fang Chaoying [Fáng Zhàoyíng] in L. Goodrich and Fang Chaoying [Fáng Zhàoyíng] (eds.), *Dictionary of Ming Biography* (New York, 1976, 2 vols.), 180–4, that Chén Dì may possibly have been influenced by the Jesuits as he had a good friend who also knew Ricci, and Ricci had made Western alphabetic writing known to Chinese scholars at the opening of the seventeenth century and maybe, informally, earlier. Chén's key study of the *Songs* was published in 1606. There is no internal evidence to support this idea, however.

126. Pulleyblank, *Lexicon* (106), 3–4.

127. H.F. Cohen, *The Scientific Revolution. A Historiographical Enquiry* (Chicago, 1994), esp. ch. 8.

128. The Sòng foot was approximately 31.6 cm; there were 5 feet per double pace [*bù*], and 360 double paces per *lí*, which was thus about 569 metres, or a little over half a kilometre. See Qiu Guangmíng, *Zhongguó lìdài dù-liàng-héng kâo* [Measures of length, capacity, and weight under successive Chinese dynasties] (Bêijing, 1992), 98.

129. Shên Gua (Sòng), *Xinjiào 'Mèngqi bîtán'* [Newly annotated 'Dream Brook Essays'], Hú Dàojìng (ed.) (Hong Kong, 1975), 322. Compare the version of this passage in Needham and Wáng Líng (58), 576. Needham and Wáng Líng translate the first two techniques as 'rectangular grid' and 'mutual inclusions', the first of which is odd and the second meaningless, and, more plausibly, suggest that at this time 'Chinese cartographers were recording compass-bearings'. There is, however, no explicit mention of a compass, and the chief hint of its presence would seem to be the use of 24 directions, which was normal on the traditional Chinese compass. See J. Needham, Wáng Líng and K. Robinson, *Science and Civilisation in China* IV. Physics and physical technology 1: Physics (Cambridge, 1962), plates between 286 and 289.

130. Aoyama Sadao, *Tô-Sô jidai no kôtsû to chishi chizu no kenkyû* [Researches on communications and topographies and maps in Táng and Sòng China] (Kyôto, 1963), pt 2.

131. J. Needham, Wáng Líng and K. Robinson, *Science and Civilisation in China*, IV: Physics (129), 220–4.

132. What would be needed is a treatment of Chinese musical theory at this period that addressed the multifaceted discussion in H.F. Cohen, *Quantifying Music. The Science of Music at the First Stage of the Scientific Revolution, 1580–1650* (Dordrecht, 1984).

133. Zhu Zàiyù, *New Theory* (10), 42.

134. The Transforming Force was not a creator but had some of the same attributes as a god.

135. Zhu Zàiyù, *New Theory* (10), 1.

136. Zhu Zàiyù, 'Jìn lüshu zòushù' [Memorial on the presentation of books on pitch], in Zhu Zàiyù (Míng), *Lülü jingyì* [The quintessential significance of the standard pitchpipes], Féng Wéncí (ed.) (annotated reprint of the 1584 edition: Bêijing, 1998), separate section, 1.

137. Zhu Zàiyù, *New Theory* (10), 2.

138. Zhu Zàiyù, *New Theory* (10), 12.

139. Zhu Zàiyù, *New Theory* (10), 45.

140. Zhu Zàiyù, *New Theory* (10), 46.

141. Zhu Zàiyù, *New Theory* (10), 44.

142. Zhu Zàiyù, *New Theory* (10), 33.

143. Zhu Zàiyù, *New Theory* (10), 18.

144. Zhu Zàiyù, *New Theory* (10), 36.

145. Zhu Zàiyù, *New Theory* (10), 41.

146. Zhu Zàiyù, *New Theory* (10), 55.

147. Zhu Zàiyù, *New Theory* (10), 18.

148. So called as its 'tone' was the proportional mean between the 'major' and 'minor' tones, or $\sqrt{((8/9) * (9/10))}$. See Cohen, *Qualifying Music* (132), 42.

149. Cohen, *Quantifying Music* (132), 41. The earliest recorded case is from 1511, but presumably it was done earlier.

150. P.A. Scholes, *The Oxford Companion to Music* (tenth ed., rev. J.O. Ward; London, 1970), 1012–17.

151. Cohen, *Quantifying Music* (132), 43.

152. This pronunciation is used for that character *zheng* ('show', 'prove', 'tax', etc.) when used in this musical sense.

153. See the diagram and explanation in Dài Niànzù, *Zhu Zàiyù – Míngdài-de kexué hé yìshù jùxing* [Zhu Zàiyù – megastar of the science and arts of the Míng dynasty] (Bêijing, 1986), 48–50.

154. See Dài Niànzù, *Zhu Zàiyü* (153), 48–54, and Needham, Wáng and Robinson, *Science and Civilisation in China* IV.1 (129), 172–3.

155. Needham, Wáng and Robinson, *Science and Civilisation in China* IV.1 (129), 185.

156. Needham, Wáng and Robinson, *Science and Civilisation in China* IV.1(129), 181–2.

157. Zhu Zàiyù, *New Theory* (10), 23.

158. Needham, Wáng and Robinson, *Science and Civilisation in China* IV.1 (129), 213, note that 'a pipe half the length of another pipe . . . does not necessarily give its octave'. The reason is that the 'effective length' of a pipe with open ends, in the sense of length of the resonating column of air, also depends on the diameter. See *ibid.*, 186, fn d.

159. From the *Shǐjì* [Records of the Grand Historian], the *Hànshu* [History of the Hàn dynasty], the *Huáinánzǐ* [The book of the Prince of Huáinán] together with the *Jinshu* [History of the Jìn dynasty] and the *Sòngshu* [History of the Liú Sòng dynasty], and the *Xù Hànshu* [Continuation of the Hàn history]. See Féng Wéncí's notes in Zhu Zàiyù, *New Theory* (10), 19. These 'methods' differed only in the numerical values they used, not in procedures.

160. Zhu Zàiyù, *New Theory* (10), 19. Emphasis added.

161. Zhu Zàiyü, *Pitchpipes* (136), 7–10.

162. Zhu Zàiyù, *Pitchpipes* (136), 9–10.

163. Dài Niànzù, *Zhu Zàiyù* (153), 226–7.

164. Dài Niànzù, *Zhu Zàiyù* (153), 296–7.

165. Dài Niànzù, *Zhu Zàiyù* (153), 122.

166. Needham, Wáng and Robinson, *Science and Civilisation in China* IV.1 (129), 218–20, and Dài Niànzù, *Zhu Zàiyù* (153), 224–5.

167. Literally 'mutual generation by intervals of 8'. There are 8 notes, with 7 semitones between them, defining a fifth.

168. Zhu Zàiyù, *Pitchpipes* (136), 'Jìn lǜshu zòushù' [Memorial on the presentation of books on pitch], separate section, 2. Also cited in Dài Niànzǔ, *Zhu Zàiyù* (153), 112.

169. The *Guózǐjian*. The translation is the standard one, following C. Hucker, 'An Index of Terms and Titles in "Governmental Organization of the Ming Dynasty"', *Harvard Journal of Asiatic Studies* 23, 1960–1, 138, but it was not a university in the full sense.

170. Zhu Zàiyù, *Pitchpipes* (136), 'Memorial on the presentation of books on pitch', 2–5.

171. Hummel, *Eminent Chinese* (7), I, 24–5.

172. Dài Niànzǔ, *Zhu Zàiyù* (153), 114–7.

173. Zhu Zàiyù (Míng), *Yuèxué xinshuo* [A new theory of musicology], in id., *Yuèlü quánshu* [Complete works on music and pitches], number 735 in the *Wànyǒu wénkù*, first series, Wáng Yúnwǔ (gen. ed.), (Shànghǎi, 1929), 5.17. Also cited in Dài Niànzǔ, *Zhu Zàiyù* (153), 225.

174. Dài Niànzǔ, *Zhu Zàiyù* (153), 235.

175. Dài Niànzǔ, *Zhu Zàiyù* (153), 121–2.

Comments on Mark Elvin

ROB ILIFFE

Mark Elvin describes a number of elements that contribute towards a better understanding of the comparative history of science and technology in China and the 'West'. He uses Alistair Crombie's notion of 'style' – developed in his 1994 *Styles of Scientific Thinking in the European Tradition* – as a means to determine absences, presences and embryological forms of what Crombie took to be the essential features of the Western achievement. For Crombie, styles were structures that 'identified certain regularities in the experience of nature which became its object of enquiry, defined the questions to be put within that subject-matter and that style, and determined the acceptable answers'. Elvin sees these as having heuristic value in defining similarities and differences between the Chinese and Western traditions and concludes that activities exemplary of virtually all of these styles were present in some form in China; in accounting for differences between early modern Europe and late Míng and Ching China, there 'is less an origin than the presence, or an absence, of an accelerating process of development of pre-existing styles of thought.' In order to deal with the broader aspects of Elvin's paper, I want to turn first to the main points of Crombie's thesis and then to a broader consideration of comparative history. Finally I will suggest that a comparative history should aim to be more sensitive to the historical richness of both traditions.

At the outset of *Styles*, Crombie envisioned his monumental account of Western science as part of 'a comparative intellectual anthropology ... comprising both its development and vicissitudes in the West and the complex diversity of forms of science arising in other cultures, with all their intricate problems of diffusion and exchange' (1: p. 5). However, this 'explicit cultural relativism', as Crombie refers to it, does not play any role in *Styles*, and indeed elsewhere Crombie goes on to claim that Western science embodies a unique commitment to truth. Instead he concentrates on the Western experience, arguing that the fundamental features of a 'scientific' world-view were originally expounded by the Greeks. His Hegelian view, in which a sort of blueprint for science unfolds or is realized over time, is accompanied by a very real commitment to a materialist history in which he describes a deep but often opaque interplay between theoretical innovation and developments in technology and instrument-making. With these 'strong' necessary conditions in place, Crombie relates, by way of his six styles, a more detailed historical narrative in which contingent processes are invoked as weaker conditions for specific developments in European natural philosophy. These include the appearance of universities, Western imperialism, Renaissance patronage of artist-engineers, capitalism, the invention of printing, new forms of mathematics and the mathematization of nature, the use of artificial situations (whether thought-experiments or experimental situations) to give information about the natural world, and the creation of new instruments and scientific journals. On a continuum, these vary from being merely highly significant factors to those that are conceivably *sine qua non* for the appearance of modern science.

Despite the evident detail in Crombie's examples, there is a legitimate question regarding the extent to which his invocation of 'styles', and indeed his generally essentialist approach to Western science, is useful for explaining the appearance of modern science. This is not the place to address this specific issue here, but by extension it does have implications for the utility of the notion of 'style' for doing comparative history. Comparative history is in principle a fruitful way of understanding differences between coherent domains of activity. On the broader historical canvas it is potentially helpful for picking out those elements that were decisive in promoting the development of Western science and technology, and more precariously, those features that were 'absent' in other cultures. Nevertheless, it is difficult to set up the two comparands in an accurate, subtle or fertile manner. In the example here, there are clearly problems concerning what could count as Western, Chinese or Islamic 'science', taking either the regions denoted by the adjectives, or the nature of practice implied by the noun as the sources of contention. Next, the issue arises as to whether Western, Chinese or Islamic attitudes towards the natural world should be taken as the 'normal' course of development, or simply the heuristic 'base' against which one or others are compared. A version of the second alternative is a more sensitive way of dealing with counterfactuals in comparative explanations, but both approaches inevitably generate accounts where some cultural 'block' or socio-economic 'trap' is invoked by which one or other – usually the non-Western culture – is prevented from developing in the way of the base society. Such explanations need not be malign (and indeed in his previous work Elvin has produced examples where they are not), although they can be, especially if the comparisons are conducted crudely through terms that derive entirely from the base society.

After this one needs to be aware of the extent to which these societies can be considered as isolated units, recognizing also the ways in which individuals or groups were active or passive recipients of information from elsewhere. One might then turn to those weaker conditions that may have been decisive in shaping the direction of Western science, or definitive in driving it forward. Such a condition may well have been the advent of universities in medieval Europe, the importance of which has been a source of contention in recent scholarship following the publication of Toby

Huff's *The Rise of Early Modern Science*. Huff claims that universities contained formally free places where unfettered debate and enquiry could take place (whereas these conditions did not exist outside Europe), and that this was decisive in fostering innovative research in astronomy and natural philosophy. Elvin himself points to the fact that there was a state university in China from the second century BCE, which flourished in the Song (960–1279 CE) setting examinations in mathematics and medicine, but which had disappeared by 1600. Whatever the case in medieval Europe, it is worth pointing out (though this does not refute Huff's point), that recent scholarship has suggested that from the late sixteenth century it was in those sites *outside* the universities, such as courts and scientific institutions, that astronomical and scientific innovation took place.

The last case shows that it is difficult for certain features to bear the explanatory load that is placed on them. In the case of the West, the tendency has been to take relatively isolated events such as the appearance of universities, or the medieval implementation of quantification, or the application of the Protestant work ethic, as decisive for the development of Western science, and by extension to make the absence of these features in non-Western cultures a definitive explanation for the success of the former. Without dismissing these particular explanations, explanations can clearly become trivial in that any aspect of Western culture could be held to be absent in any counterpart; descriptions of these situations cannot simply be invoked as explanations. In my view the only viable procedure is to find loose mid-level features that are neither of the ahistorical nature or monumental scale of 'styles', nor too local. Having said that, there are clearly strong necessary conditions that were of pre-eminence in Western science, such as the use within natural philosophy of instruments such as lenses and chemical apparatus, the capacity to represent parts of the world mathematically, and the existence of institutions and communication networks through which individuals could access and operate on critical masses of information.

All this leads to Elvin's fascinating examples. By way of analysing the 'experimental' style, he describes the difference between straightforward observation (of dragons and snowflakes) and the more complex situation in moral epidemiology by which a large situation was turned into a sort of controlled experiment. In other fields, such as in medicine and metallurgy, complex analysis of situations clearly embodies an understanding of a basic form of the experimental method, although as Elvin himself suggests, there is a sense in which 'every skilled cook, gardener, or craftsman in China was in one dimension – but usually only one – an experimenter who strove to achieve … effective repeatable control of the causal inputs and outputs.' This attitude, while lacking something of an ethical dimension, is wonderfully exemplified by the instructions on how to test potent toxins on one's relatives.

As for modelling, Elvin distinguishes between devices that simply produce the appearance of a process, and those that offer a causal account of nature by recreating macro-processes on a smaller scale. Within late medieval and early-modern Western astronomy debate about whether Ptolemaic or other models described the real motions of the heavens, and whether these models also represented the causal mechanisms of the heavens was endemic. Complex physical models of the heavens existed in China as early as the second century CE, although in the example of Zhang Héng cited by Elvin it is not clear whether the mechanism of the model was intended to mimic the actual working of the heavens. For the 'taxonomic' style, the example of Lî Shízhen's *Pharmacopoeia* is yet more proof of the degree of the relatively advanced procedures of which Chinese scholars were capable. Li's penchant for comparison of textual authorities, coupled with the occasional consideration of his own observations, resembles the practices of humanist naturalists in the early-modern period. As for probabilistic thinking (or rather reasoning), Elvin points out that *The Book of Changes* embodies a sophisticated understanding of probability. Nevertheless, in the remarkable case of 'sale by gambling', he points out that despite the fact that shopkeepers would have had to have a sound feeling for the odds as a basis for continuing their existence, 'there is no reason to believe they knew how these odds could be calculated theoretically'.

Regarding the presence of historical derivation in Chinese thought, Gù Yánwû's extraordinary mid-seventeenth-century *Theory of Reading Pronunciations* shows an attitude to the past that, as Elvin says, is somewhat similar to the Renaissance vogue for a lost originary truth or *prisca sapientia*. Interestingly, this research was also based on a rigorous and collective comparison of different textual sources. Finally, although this does not correspond specifically to a Crombeian 'style', Elvin usefully describes the research undertaken by the sixteenth-century Ming prince Zhu Záiyù on the proper size for a pitch-pipe required to produce a truly aetherial harmony. Zhu deployed empirical observation and experiment, and recognized that (in his words) 'Pitch is generated by numbers. If the numbers are correct, the pitches will all be in accord with each other.' Zhu's work, which involved a sophisticated application of mathematics to one feature of the natural world, was printed in 1595 but his doctrines were largely ignored until attacked and then defended in the middle of the eighteenth century. The delay was probably due to internal political issues, and by way of conclusion Elvin rightly comments that differences between the cases of Zhu and Galileo 'need subtle pondering'.

Pace Elvin's positive comments about the usefulness of Crombie's 'styles', I take it to be a key achievement of his essay to have shown the limits of these concepts for comparing Western and Chinese approaches to nature. Even if the concept of styles is helpful for analysing Western science, using them for comparative purposes must inevitably render Chinese examples incomplete analogues of their Western counterparts. More generally it is of course difficult to avoid couching Chinese achievements in Western terms for a Western audience. Despite the signal success of Needham and his colleagues in determining and then drawing attention to the sophistication of Chinese achievements, the need to stress these attainments as equivalent to or in advance of Western work often resulted in a lack of sensitivity to the Chinese contexts. Elvin reminds readers once again just how significant were the achievements of premodern Chinese science and technology, while he does not allow his use of 'styles' to twist the meaning of the original text. In an important passage he writes that the 'weak but seemingly unfruitful resemblance[s]' between China and the early-modern West 'make the positing of sharp contrasts untenable'. If one moves away from these sharp contrasts, it seems to me one can move away from the constraints imposed by abstract notions such as 'styles' to a more historically sensitive account of both the Chinese and Western cases.

I would argue that the vast majority of scholars working on comparative history need a much deeper grasp of the contexts of what Chinese scholars and others were doing, as well as of the history of Western science. Deploying more sensitive categories may also lead to a situation where we need no longer see isolated pockets of Chinese activity as *merely* 'seeds' that fail to grow [into Western science]. I use the qualifier in the previous sentence because some sort of basis for comparison is crucial, both to ask what was special about the West, and also to understand why Chinese science and technology was the way it was. If comparative history is to become more anthropological – recognizing the real *differences* between Western and non-Western cultures through presences in the latter – then one might be able to forge approaches that do not skew the analysis by employing overly precise Western categories. As a means to achieving this, I look forward to reading comparative histories that employ Chinese terms and concepts.

Inside Newcomen's Fire Engine, or: The Scientific Revolution and the Rise of the Modern World

H. FLORIS COHEN

A TEXT TO START FROM

A most useful text for us to take as our point of departure is a 3-page piece 'The Road to Riches' which appeared at a prominent place in the special 'Millennium' issue of *The Economist*.[1]

The leading idea of that special issue was a highly felicitous one, namely, that nobody in their right senses looking a thousand years ahead from the vantage point of the previous millennium in the year 1000 AD could possibly have foreseen either the contours of our modern world of the year 2000 or, *a fortiori*, that Western Europe is where the big leap towards that modern world of ours occurred. There were a number of advanced civilizations at that time, by and large on a par mutually, with Europe appearing as rather a latecomer among them; certainly, at first sight it is a profound enigma why, of all civilizations then around, it was this latecomer Europe that, some nine centuries later, was the one to make the unexpected leap. But the greatest thing about *The Economist*'s millennium issue and that one piece, in particular, is that, to the extent that solutions to the enigma are tentatively being put forward, these are neither stubbornly monocausal nor learned elaborations upon simple-minded one-liners about Westerners being more logical or having better genes or some other allegedly built-in, timeless superiority. By the same token, these answers are equally far removed from the other extreme of turning desperately *non*causal, as if where the leap did occur is just a matter of chance unfit by definition for any deeper historical analysis.

The 3-page piece 'The Road to Riches', in its first-rate journalism, then, provides a particularly good example of a sophisticated approach in which one cause, no cause, and cheap sloganizing are all being shunned in favour of some lucidly and succinctly rendered fruits of genuine, also wide-ranging and boldly comparative historical analysis. It takes as its starting-point the idea that, with modern science and technology providing so much of a motor drive for *present-day* economic growth and social change, it seems rather obvious to seek the motor that originally drove the European leap likewise in science-based technology with its meanwhile proven capacity for social change of the most drastic kind.

However plausible that anwer may sound – so the argument in *The Economist* goes on – though unlikely to be entirely wrong it fails to satisfy on at least three significant counts:

1. Up to Edison's times around 1875, the huge majority of industry-promoting inventions, such as the spinning-jenny at an early stage, owed virtually nothing to science.
2. If the rapid advances in science wrought in the seventeenth century by pioneers like Galileo or Huygens or Hooke, who were all so keen upon practically exploitable application of their finds, had indeed been decisive, whence, then, the passage of at least a hundred years to come until industrialization even began to get going?

These two points lead to an interim conclusion, which reads: 'The link between science and technology is subtler than you might think.' But a further objection to any straightforwardly causal connection between European science and European industrialization is raised in addition:

3. Other advanced civilizations (notably China before, under the Qing, it lost its lead for good) had experienced periods of brilliantly flourishing science and/or technology, coming to the brink of industrialization in at least the Chinese case, yet never getting beyond it.

The net conclusion drawn in the article at this point is that 'science and technology, in short, can get far and then stop', so that we find the truly crucial question to run thus: 'What happened in Western Europe in the seventeenth and early eighteenth centuries that failed to happen in the Western Europe of antiquity, or in China after 1400, or in the Islamic world after 1200?'

At this point of high suspense the article makes a turn that does not cease to amaze me. Without any pause it goes on to tell us that 'the question is ferociously debated by economic historians', thus leading on the final page, once again without any pause or interruption, to a once again quite knowledgeable discussion of the 'three broad and overlapping things [that], between them, made the difference: values, politics and economic institutions.' Nowhere in this concluding section does either science or technology, let alone the 'subtle link between them', make its reappearance. True, it follows from objection (3), in particular, that they cannot do so as *independent* variables. But the coherence of the overall argument requires in no way their absence as *dependent* variables. Ongoing silence on the subject from that point onwards is the more surprising as now (if we make the conclusion thus drawn face the consideration with which only three pages earlier the piece took off) the odd paradox ensues that, whereas modern science and technology evidently add very substantially to all that drives our present-day, modern society forward, nonetheless modern science and technology, in being neither 'values' nor 'politics' nor 'economic institutions', nor apparently being among the major phenomena conditioned by any of these three, had nothing at all to do with how modern society originally came into being. Not a very plausible state of historical affairs, to say the least. Why, then, call attention in a serious historical journal to a mere journalist's sloppy reasoning?

One reason for doing so is that we had better not think of journalists in terms of 'mere journalists'. Not only do the contents of *The Economist* reach an immensely larger and politically more influential readership than any scholarly attempt to set its conclusions right may even begin to hope for. But we should also realize that what we have here is really journalism at its very best, not only for the reasons I gave in my second and third paragraphs, but also in that it reflects, and ably sums up, a good deal of current, scholarly thinking on the subject.

Whose thinking? Well, chiefly economic historians' thinking. The curious dichotomy between economic history and history of science and technology displayed in *The Economist* piece is nothing but the reflection of a dichotomy that reigns in the world of scholarship. In my *The Scientific Revolution. A Historiographical Inquiry* of 1994 I complained that historians of science have been in the habit of treating the emergence of modern science as 'a secret treasure'.[2] By this I meant to say that, although indeed historians of science have illuminatingly conceptualized the emergence of modern science in seventeenth-century Europe and elucidated what was so uniquely novel about the kind of science that did emerge, with few exceptions they have failed to link this up other than in slogans with what the event has meant for the coming into being of our modern world. Indeed, it is in good part due to how the common run of historians of science has chosen to ignore that world-historical problem and to marginalize the few who have not, that the problem, which really requires the expertise of just about all varieties of historians (those of politics or of religion included), has become so much the domain of economic historians alone. I certainly do not want to begrudge them their concern; to the contrary, it is altogether a great boon that at least in one speciality (also the one where sophisticated, analytical-comparative thinking appears to have advanced to far greater heights than in any other branch of history) the problem has remained alive. Further, an increasing number of chiefly economic historians has in past decades been crossing these borders (the names of Landes, O'Brien, Mokyr, Goldstone, whatever their mutual differences, come to mind at once). Still, to show what is tenable and what not about the three specific objections leading in 'The Road to Riches' to the exclusion of modern science and technology from the story of how the world got modern and rich, which objections are really not sloppy at all yet not good enough either, requires perhaps a bit more familiarity with findings made by historians of science and technology, even though these have rarely made their findings sufficiently general or exploited them for their world-historical implications.

My aim in the present paper, then, is to confront all three objections advanced in that *Economist* piece, doing so from the point of view of a historian of science concerned to understand of the Scientific Revolution in seventeenth-century Europe (a) its major components (dynamics and productive interactions included); (b) how it could come about at all; (c) how it is that it did so in Europe, not elsewhere; (d) what it meant for traditional craftsmanship in the shorter and longer run; and (e) what it has contributed to the coming into being of our modern world.[3] In discussing, as I presently shall, pertinent issues raised by our three *Economist* objections I take the harnessing of steam power to be indispensable for arriving at a fair judgement of the piece as a whole, in that steam helps us see

- why objection (1), albeit not just wrong, is decisively exaggerated;
- how we had better rephrase, and then go ahead to answer, the very good question posed in objection (2);
- why objection (3) is untenable, with consequences.

The pertinence of steam power for his argument has not, to be sure, gone quite ignored by the (anonymous, as is *The Economist*'s wont) author of 'The Road to Riches'. In regard of his objection (1), he does mention (albeit without any further explication or discussion of consequences for his own argument) the steam engine as one exceptional piece of science-based technology prior to Edison's times. In regard of his objection (3), he very strongly suggests (without quite committing himself) that in China even earlier than in Europe all the knowledge about atmospheric pressure that one might need to build a steam engine had already been discovered without that knowledge actually being tapped for going ahead and constructing such a machine. Only in regard of (2), on the application-oriented science of Galileo and Huygens and Hooke and its failure for at least a century actually to be applied, does the author fail to mention steam power as an example, even though well he might have; as we shall see.

    In order to lay bare what I regard as the pivotal issue on which the *Economist* piece goes wrong in its effective downplaying of what all that happened in science in seventeenth-century Europe truly meant for the process of industrialization, albeit indeed one century later, let me take you inside Newcomen's engine. Although in so doing I aim to make you consider it afresh, in my brief explanation of how it works I claim no novelty whatsoever – it is all there in the literature on the history of steam technology,[4] only the message of that literature does not seem always to get out to all whom it concerns.

NEWCOMEN'S ENGINE, FROM THE OUTSIDE IN

To ease your entrance, I start from the outside (see Figure 1).[5] Note that the original caption does not speak of a *steam* engine, but rather reads 'The ENGINE for Raising Water (with a power made) by Fire'. Also note that, thus looked at from the outside, there appears to be nothing in its construction that seems principally out of reach for any comparably daring inventor in any comparably advanced civilization of the premodern world.

Figure 1   Newcomen engine. Beighton's engraving of a Newcomen engine, 1717.

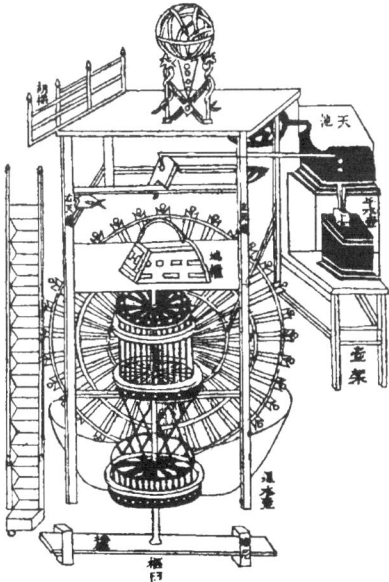

Figure 2   Su Sung's water clock. General view of the works of Su Sung's clock tower.

Figure 2, for example, is a wood-block engraving of Su Sung's huge water-clock regulating the course of the planetarium-like armillary sphere on top of it.[6] This invention, rediscovered in the 1950s by Joseph Needham, dates from the late eleventh century CE, from Northern Sung times, that is. Looking at all this from the outside, no good reason presents itself for why a civilization capable of both inventing and constructing so sophisticated a piece of machinery could not in time have produced a fire-engine like Newcomen's or even the kind of steam engine Watt later transformed it into.

We now move closer inside.

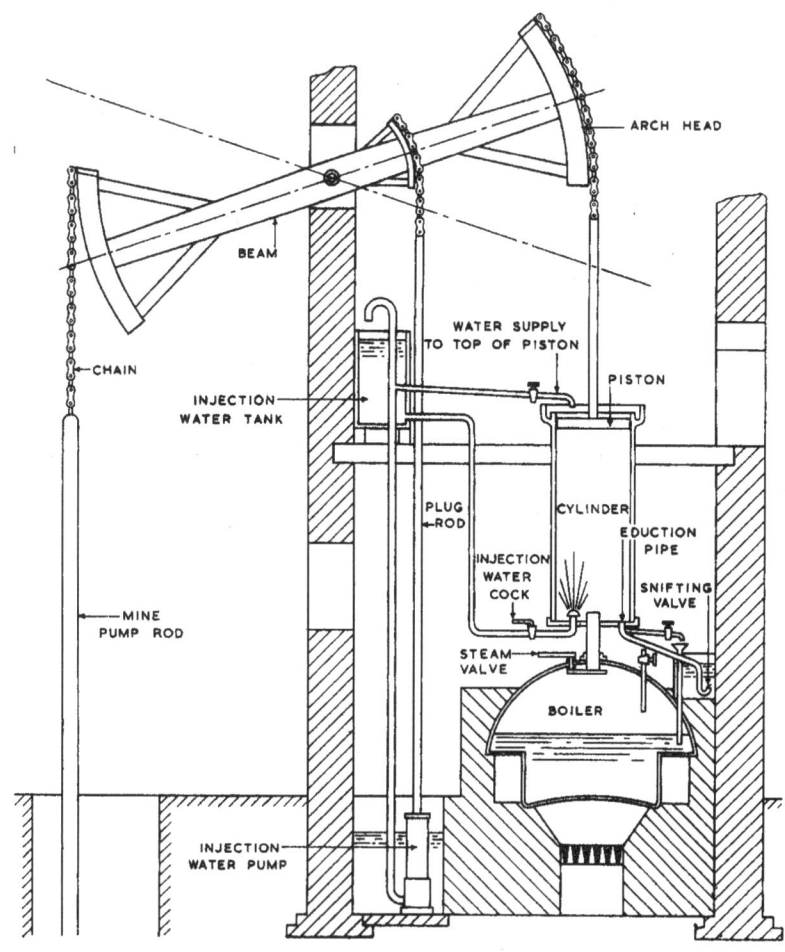

Figure 3 Diagram of Newcomen engine. Diagram of Newcomen's atmospheric engine, 1712.

In Figure 3,[7] note first the big balance beam, with the mine pump rod hanging from a chain on the one side, and a piston on the other. The rod enables suspended buckets to pump water out of Britain's coal mines, which is what the machine was made for. The piston fits closely into a cylinder which can be filled with steam from the boiler below, in a cycle regulated by the steam-valve in between. Also suspended from the balance beam is a small pump, allowing the cold mine-water it raises to be used for being sprayed into the cylinder. This likewise happens in a cycle, regulated in this case by the injection water cock in between. Finally, water leaves the cylinder through an eduction pipe and air leaves it through a 'snifting valve'.

In order to grasp how this assembly causes the balance beam to make the mine pump rod go up and down and thus empty the mine of water, we go further inside the engine, and seek to understand what the picture can no longer show. The effect of spraying cold water into a cylinder filled with steam is to condense the steam, thus leaving a vacuum inside, so that the atmosphere presses down the piston, thus drawing the mine pump rod up. With the injection water cock shut off and the steam-valve opened, fresh steam enters the cylinder, the piston is pressed upwards, and the process can repeat itself, up to some 15 cycles per minute in practice for more than a century upon the engine's invention around 1710.

The invention was not a wholly novel one. Leaving aside Savery's rival engine, which for clear-cut reasons proved far less successful, there is no reasonable doubt that, whether directly or indirectly, Newcomen must have been aware of a sketch communicated by Denis Papin to the Royal Society and published in its *Philosophical Transactions*. Let us consider now three distinct features of Newcomen's invention. One is your skilled craftsman's improving upon a scholar's proposal – a practice-oriented scholar to be sure, yet a scholar. This feature stands expressed primarily in Newcomen's introducing a separate boiler rather than, as Papin had proposed, putting a fire under the cylinder but the water in it. A second feature is Newcomen's genius as an inventor, which stands expressed in the presence of the snifting valve and, above all, in the automated snap-action of the injection cock and the steam-valve. The third feature is what I am most concerned with for now. It is the unseen heart of the machine, the process of deliberately creating a void so as to enable atmospheric pressure, rather than horses or your fellow man, to do work for you. What we have here is something transcending in a principal manner mere premodern craft ingenuity, however admirably ingenious at its best. I have just called Newcomen a genius. But both previously and later, both here and elsewhere the world has known technicians blessed with equal gifts. What, then, is so special about this fire-engine by Newcomen to make it the first product of technical ingenuity decisively different from, say, Ktesibios' or *a fortiori* Su Sung's water clock, from Chang Hêng's seismograph, from al-Razi's distillation vessels?

REVOLUTIONARY SCIENCE IN THE ENGINE

My answer to this question is fourfold. It is that at the heart of the machine we find the incarnation of a scientific idea – an idea made flesh; an idea explicitly formulated and deliberately applied. It is, next, that that scientific idea, which is about the possibility of void space and of experimental proof for the effective voidness of certain spaces, was entirely unheard-of, without any precedent anywhere. It is, further, that that unprecedented scientific idea runs counter to any common-sense conception of nature. And it is, finally and most importantly in my view, that that novel scientific idea is not just an isolated, bright idea such as may come up as a matter of course in some chain of ongoing, regular scientific advance, but is just one sample of something principally novel occurring at the time. What is principally novel about it, is that the body of ideas and practices out of which the recognition, creation and effective proof of the void came forward alongside much, much else, comes down to a well-prepared and nonetheless quite radically novel mode of thought about nature as such.

This mode of thought is that of modern science as it first came up in a – to us nowadays – *grosso modo* recognizable way in seventeenth-century Europe. This mode of thought about nature has in its broad outlines become quite routine to us nowadays; at the time, however, it looked extremely odd, even menacing, in that it went flat against a whole range of conceptions shared across cultures the entire premodern world over. For the first time, nature now began to be addressed on a principally other plane than that of either one-sided empiricism – with facts taken from everyday sense-perception and sheer common sense and then arranged in orderly patterns on the same level of sense-perception and common sense – or of an even more one-sided intellectualism, with everyday observations of nature pressed into one or another set of all-encompassing, philosophical first principles deduced *a priori*. In short, a thoroughly *counter-intuitive* mode of nature-knowledge at a plane of understanding unprecedentedly intermediate between abstract philosophy and plainly concrete common-sense observation came up in seventeenth-century Europe, and this mode of nature-knowledge was both immensely more *reliable* and, at least potentially, more *useful* than any previous mode of acquiring knowledge about nature.

That vastly increased *reliability* about the mode of nature-knowledge that seventeenth-century Europe began to pioneer, and the very principal reasons for it, are at the heart of what our *Economist* piece overlooked (thus faithfully conforming indeed to the great majority of current debates over what made Europe stand apart).[8] That vastly increased *usefulness*, as our *Economist* piece noted well, was apparent to many a proponent of the new science right from the start. The trouble, as it perceptively noted as well, was that useful application of a whole range of novel insights kept eluding them in almost every instance. This is true of nearly all those promising plans put forward over the entire period of the Scientific Revolution from Galileo up to Newton. Be it the harnessing of atmospheric pressure by making water boil on the bottom of a cylinder and then allowing it to condense by putting the fire underneath out (Papin, elaborating upon Huygens, von Guericke, Pascal, Torricelli and Galileo), or to regulate water streams (Castelli, elaborating upon Galileo), or to make the deaf hear again (Hooke, elaborating upon Bacon), or even to produce artificial manure out of wood juice (Glauber, elaborating upon Paracelsus), it all came to naught.

Why did it all come to naught? Here again the *Economist* piece is quite right: the link between science and technology is subtler than we tend to think. More than that, prior to the seventeenth-century Scientific Revolution there was scarcely any link at all. The kind of nature-knowledge that we can see by hindsight to have been applied in practical tools and machinery was intuited; it was neither derived from, nor supported by, any available body of nature-knowledge. We can see how the construction of certain stops in late medieval organs brilliantly exploited properties of the so-called overtones, or harmonics – the very beginnings of acoustical knowledge of those harmonics, i.e., their one-by-one identification as such, took place more than a century later, in the ears and the quiet monastery of Father Marin Mersenne during that selfsame Scientific Revolution.

To find out how the customary chasm between nature-knowledge and the crafts gave way to the kind of science-based technology we tend to take for granted nowadays is among the most neglected empirical issues central to the Scientific Revolution. I shall now report on conclusions I have drawn from a survey set up with a view to identifying what it was that left those bold claims one almost hears ringing down the seventeenth century so fruitless for the time being. What hurdles were being encountered in practice, and what enabled scholars and craftsmen of later times to overcome them?[29] In the survey that follows my focus is specifically on seventeenth-century efforts to make *mathematical* science practicable.[10]

BETWEEN THE BACONIAN DREAM AND ITS REALIZATION

What is a machine? During the European Renaissance, prior to the birth of modern science, from the manner artisans used to handle machines like, for example, saw-mills or waterwheels, we can deduce that they regarded these as, so to say, composites of individual parts to which, so as to enhance their effect, one could add further parts at will. As one truly precocious exception, Leonardo da Vinci alone began to perceive, in the privacy of his handwritten notes, that machines are rather assemblies coherently made up of an identifiably circumscribed number of constructive elements. His desire to optimize the effect of machine tools, guided by an awareness that what one gains one way is necessarily lost another, led him toward no less exceptional ranges of experimental researches into possible chances for minimizing friction, in particular.[11] About a century later, Galileo unwittingly took his departure in the very same fundamental insight into the nature of machine tools, so profoundly at variance with most craftsmen's received wisdom on the subject. However, instead of

Leonardo's early, fact-finding experimentation, Galileo characteristically went on to pioneer the other, mathematical component of the beginnings of a theory of optimal engineering practice, by arguing that all machines are jointly reducible in the end to the law of the lever. Here is how one rare historian of technology-*and*-science, the late Donald Cardwell, expressed the major practical change thus wrought in theory:

> How could one possibly compare a fulling stocks with a corn mill, a saw mill with a blast furnace, a mine pump with a pump for supplying a mansion with water? All of them did entirely different things and the only common question to ask was whether each machine served its purpose well. The answer ... could only be normative: it was a good machine or it was not. But according to Galileo's arguments all machines, no matter what purposes they serve, have the common function of transmitting and applying 'force' or power as efficiently as possible and, moreover, the performance of machines can be quantified, for ideally the product of the driving 'force' and its velocity equals the product of the load multiplied by its velocity.[12]

Here we watch revealed at its very clearest the (on the longer term) world-shaking significance of the domain of common-sense and rule-of-thumb driven craftsmanship being drawn into the counterintuitive, all-disturb-ances-removed, ideal-case domain of modern, mathematical science which Galileo pioneered and which has remained the core of modern science ever since. Single-handedly, Galileo showed that (to quote Cardwell once again)

> a machine croaking and groaning under its load is not, in spite of appearances, necessarily doing the most work, [in that Galileo] saw that the inequality between the equilibrium force and the force required to set the machine in motion was not a principle of Nature; it is merely a consequence of the imperfection of all machines. A good machine will move at a uniform, unchanging velocity when the force and the load, including residual friction, are in equilibrium. A perfect machine, free from friction, distortion and other defects, will accelerate (slowly if loaded) under an extremely small force until it reaches an infinite velocity.[13]

Now it was one thing to set common-sense convictions right in theory; still quite something else to draw practical benefits from these and related insights – to make craftsmanship enter the domain of mathematical science was hardly tantamount to turning it by the same token into science-based technology. Not Galileo or his disciples, but Edme Mariotte in the second half of the seventeenth century stood at the head of a range of efforts undertaken with increasing frequency over the eighteenth century actually to calculate (however often outcomes proved mistaken at first) the work capacity of, notably, waterwheels and, later, engines using steam for their source of power.

This, then, is my theme here – the vast gap customarily obtaining between premodern science and premodern craftsmanship; such chances for bridging that gap as were perceived right from the start in the early seventeenth century to be inherent in modern, Galilean, mathematical-experimental science; and, above all, the question of how realistic those expectations then proved to be in the shorter as well as in the longer term.

Let me first pursue this theme of mathematical science and the crafts a little further. As compared to the two civilizations where the mathematical handling of some scattered pieces of the natural world had flourished before, that is, in Hellenistic civilization, which pioneered it, and in Islamic civilization, which adopted and enriched the approach – as compared with these two, Europe went a good deal further in exploring interfaces with the crafts. This was a highly unusual undertaking. But for some legendary claims about Archimedes, and some actual applications of the law of the lever in pulleys and screws as described by Hero, the highly intellectualist pursuit of knowledge of nature and the trial-and-error construction and improvement of such tools as (to mention two particularly advanced ones) eyeglasses or mechanical clocks went ahead in virtually watertight separation from one another. In this respect, the direct application of geometrical and/or arithmetical rules in linear perspective, in navigation and map-making, and in several military arts that took place in the European Renaissance, on however modest a scale in retrospect, was without world-historical precedent. But we must consider that, by 1600, mathematicians had already come close to exhausting what they could usefully contribute to the creation of an illusion of painted space, to the outlay of fortresses, and to the determination of geographical latitude and the making of maps. So this might well have been the end of an interesting, relatively short-lived episode of collaboration between mathematicians and professional painters, soldiers, sailors and map-makers. But at this very point of imminent exhaustion a new vista arose from Galileo's conviction that our empirical world is at bottom mathematical, joined to his actually demonstrating how empirical phenomena like free fall and projectile motion can indeed be handled using mathematics. As he was very well aware, a wholesale transformation of traditional craftsmanship had thus come within – in his view, very close – reach. Bracketing the often grandiose promises that went with such expectations at the time, I have surveyed what achievements were actually attained in the dozen or so areas of craft activity actually subjected to some degree of mathematical treatment in course of the seventeenth century. How did the newly realist bent of gaining knowledge of nature the mathematical way affect the ongoing pace of events in the realm of the crafts?

For an answer, I must first stress, as an utterly basic point, that we should *not* indulge our natural temptation to read the resounding successes of science-based technology that are so familiar to us back into those first probings undertaken in Europe prior to the advent of the very mode of science that was to make those successes possible in the first place. Nor should we assume *a priori* that Galilean mathematical science, which indeed was to make so much of the difference in the longer term, necessarily began to make that difference all at once. In fact, my survey of literature devoted to examining a variety of crafts in seventeenth-century Europe has led me to conclude that traditional craftsmanship kept prevailing throughout the Scientific Revolution because mathematical precision was either hardly or not at all attained, or was attained indeed but proved practically irrelevant for the time being. The crafts most decisive for everyday life (those responsible for food, clothing, shelter) remained in their traditional, trial-and-error state, whereas, without exception, often quite sophisticated mathematical thinking directed at the solution of more or less pressing practical issues really proved impracticable for the time being. Whether we find ourselves dealing with musical tempering; with the mathematical improvement of windmills used for draining; with the taming of water streams in the Po delta and the Venetian lagoon; with efforts to enhance shooting range and improve shooting accuracy the mathematical way; with estimating and comparing the strength of materials; with the measurement of geographical longitude on board ship; or with the determination of machine efficacy with which I started my discussion of these matters – for all the sophistication of the mathematical approaches advocated and most often tried out as well over the seventeenth century, in not one of the various crafts involved did practitioners feel by such efforts called upon to alter their customary ways of proceeding.

How could this be so? How is it that when, by century's end, the Scientific Revolution came to some provisional point of culmination in Newton's *Principia* and *Opticks*, the gap between mathematical science and the crafts yawned almost as widely as it had at the onset of the Revolution, about a century earlier?

One principal reason is that, from a contemporary point of view, there was scarcely ground for suspecting at the outset that the gap was so wide in the first place. A marvelous new tool of, in principle, great generality had now become available, the subjection of the empirical world to mathematical rule and order; why should the empirical world of craftsmanship fail smoothly to fall into line? Although it does not appear that, at the time, the issue was ever posed in anything like so clear-cut a manner as we can pose it in retrospect, we may well regard the Scientific Revolution as being marked by a voyage of slowly yet surely advancing discovery and that is why the gap could not be overcome by jumping over it in one big mathematical leap, but required a good deal of patient, laborious bridge-building.

What bridges, then, and who had to build them, craftsmen or mathematical scientists?

This differed considerably from case to case. Craftsmen's virtually innate conservatism, their aptness to leave things that work intact and to seek solutions for upcoming practical problems in the general direction of what has already proved workable before, is sometimes quite functional, sometimes less so. It is very hard, often impossible, however, to decide in advance when we are dealing with experience-based, sound judgement, and when with no less experience-based prejudice. Werckmeister, the church organist, can in retrospect be seen to have done well to ignore mathematical scientists and their tendency to prefer both mathematical and musical elegance over practicality, whereas Galileo's analysis of machine efficacy, which likewise went against craftsmen's intuition, and was likewise devoid of possible practical consequences at least for the time being, was in due time to alter craft practice almost beyond recognition. It is hard for us not to sympathize with the predicament of those cardinals who, in the early 1690s in a committee on the river Reno in the Po delta, had to choose between the competing claims of Guglielmini, the somewhat haughty mathematical modeller, and of his Jesuit opponents and their down-to-earth experiential bent in the tradition of the great Renaissance artisans. But we cannot but sympathize as well with Huygens' complaint, in a report about the performance of his clocks on the route back from the Indies, how much their overseers had suffered 'from the crew's frequent scolding and mockery of this effort to measure longitude in a new way' which, as Huygens rightly foresaw, was in due time to save the lives of so many sailors.[14]

But the survey I have made also leads me to draw some conclusions valid for all these varied cases.

First, the degree of urgency of the particular practical problem for which mathematics was adduced as a solution made no difference at all for the outcome. Whether perceived at the time as pressing (notably, water management and the determination of longitude on board ship) or not (in particular, the efficacy of machines and the strength of materials), none of these problems had even come close to resolution in practice by the century's end.

Demand, then, did not sufficiently foster resolution; but neither did craftsmen's conservatism necessarily stand athwart it. In many a case the very initiative came from craftsmen; even if not, when plausible modes of change in a new direction were put forward by mathematical scientists (in perspective, fortress building, navigation, map-making), craftsmen ready to go along invariably came forward. From whatever, mostly status-related sources their conservatism might spring in addition to the inherent one already adduced, the more open-minded among them could overcome it; what, then, did prevent such plausible modes from presenting themselves in all other cases?

The following impediments to practical success for the one big mathematical leap can retrospectively be identified:

a. The weightiest impediment of all rested in a severe underestimation of the real world's messiness. For craft after craft it appeared in the course of the Scientific Revolution that, beside determinants taken up indeed in the mathematical model, there were others of at least equal significance, and also many second-order effects, which somehow had to be brought under mathematical rule as well.
b. The mathematization of second-order effects, in particular, required an

ability to handle non-uniformly varying magnitudes which the Euclidean doctrine of ratios was inherently incapable of satisfying but which was to be yielded by the calculus.

c. Even if using Euclidean means only, craftsmen's ability and/or readiness to grasp the esoteric language of mathematics was very limited, nor were there as yet enough incentives for the drastic educational reform required to make fruitful communication between the craftsman and the mathematical scientist a matter of more than just rare good luck.

d. Such communication was further impeded by social distance. Many a mathematical scientist stemmed from the lower or higher nobility or had to adopt its standards in any case, and thus tended to regard the equal footing really required for fruitful exchange as beneath his dignity.

Due to at least these four retrospectively distinguishable barriers, then, the crafts stood to gain very little from those numerous efforts undertaken with a view to their scientific improvement (I mention only in passing that, vice versa, mathematical science did benefit greatly from the experiences thus undergone, which in most cases centered around the utterly novel practice of mathematical modelling).

To return to our gap, we do find in course of the century some budding recognition of this state of affairs, as well as some early efforts to overcome it. The first recognizable cluster of crafts/mathematics interaction consisted of Galileo and his Italian disciples, and was marked throughout by their aptness to stick very much to one-sided, mathematical idealization in the sometimes more, sometimes less complacent expectation (fed by their still brand new conception of the world as inherently mathematical) that experiment would confirm it or, if not, that it could be conveniently reasoned away. In the second half of the seventeenth century, in the Académie Royale des Sciences, Huygens and Mariotte, in particular, began, also mathematically but in a more open-minded experimental vein, to take the world's messiness into account and to explore second-order regularities. Finally, whereas contemporary concerns of the Royal Society with how to make the new science impinge on craftsmanship took a distinctly Baconian, hence, non-mathematical direction overall, in book II of Newton's *Principia* we begin to discern a first glimpse of what the calculus, once applied to practical issues, might in due time be able to accomplish.

The removal of impediments (a) and (b), thus instructively explored by mathematical scientists in course of the seventeenth century, was to be pursued with ever increasing zest and refinement over the next, producing one mathematical model after another of sufficient sophistication to become of real practical value. In a few cases mathematical scientists (prefigured in this regard, too, by Huygens, by Mariotte, and also by Newton in the sole case of his reflecting telescope) turned themselves into something resembling craftsmen, yet most often the emergence in course of the eighteenth century of craftsmen of a thus far wholly unknown type like Parent or Smeaton was to prove decisive in removing, or at least softening, impediments (c) and (d).

In terms of our *Economist* piece these conclusions mean that its objection (2), about the great pioneers of seventeenth-century science fostering practical application of their work in vain, is fundamentally sound. True, I have arrived at my conclusions solely for *mathematical-experimental* science, since it provides the clearest-cut case. But non-mathematical yet otherwise similar efforts undertaken in Baconian circles to make their revolutionary, fact-finding experimentalism such as had equally arisen in revolutionary fashion in the seventeenth century serviceable in practice, display very much the same pattern. I already mentioned in passing Hooke's vain hopes to make the deaf hear again, and Glauber's schemes for turning wood juice into manure. It has further been shown by Westfall how Hooke's naturally pre-Newtonian dynamical insights, in particular, stood in the way of the successful resolution of certain other practical problems he had posed himself and believed he could solve with science (e.g., how to make a lamp burn steadily; or the optimal way to trim sails).[15] And of course I have already broached the case of atmospheric pressure, which I shall now elaborate a little further since, early in the eighteenth century, it was the very first to undergo a decisive twist.

The story of what eventually (unplanned and unforeseen) became the high-pressure steam engine is well known to take its departure in Galileo's becoming aware of craft experiences with the suction pump, with the apparent impossibility to pump water to greater heights than *c.* 10 metres leading him to a radically novel, non-philosophical, newly-scientific manner of thinking about the void – not as an issue to be decided *a priori* from first principles, but *a posteriori* through mathematical theorizing *cum* experimental testing. His disciples then pursue this further with mercury and persuade themselves of the void in Torricellian space, and of the column of heavy air balancing the mercury column below the void space, with Pascal elaborating and experimentally confirming all this. In a further, more directly practice-oriented way von Guericke seeks to measure and exploit the pressure exerted by the atmosphere, which Boyle also experiments with. Huygens sketches a design for an engine hopefully capable of the cyclical re-creation of a void space using gun powder at the bottom of a cylinder; his assistant Papin replaces gun powder with boiling water and its cyclical condensation. So far events fully fit our pattern for both mathematical-experimental and Baconian-experimental science – with Papin's device we have now in hand the theoretical solution to a practical problem, with, however, the practical solution characteristically way out of sight and, hence, current craft practice in the draining of mines so far quite unaffected. But then our decisive twist sets in, for which Papin's crossing the Channel may stand as a neat symbol. Despite numerous, not too thoroughly thought-through schemes for the improvement of daily life finding enthusiastic, almost weekly discussion in the Baconian circles of the Royal Society, very little of which ever came to anything; despite all those fruitless undertakings unforgettably caricatured in Gulliver's Voyage to Lagado, it is on the European Continent mostly that we have encountered the (if I may call it thus) theory of practice. The onset of its actual, practicable realization in the course of the eighteenth century, however, proved a feat in large measure British. Why?

A BRITISH EXCEPTION

To answer that question, let us take an indirect approach, and focus briefly upon Cardwell's enlightening comparison between how, in the 1760s, Watt and Smeaton, respectively, approached the Newcomen fire engine.[16] Personal traits are surely responsible to some extent for their differences – Smeaton a highly gifted man capable of reaching the outer bounds of what a given line of attack still can yield; Watt a genius endowed with the rare gifts of 'lateral thinking' and of saturating oneself with a problem largely of one's own making and then hitting upon the (only in retrospect obvious) solution. But their respective intellectual prehistories also count. Smeaton, as noted, had mastered the calculus, which was quite irrelevant to the job of improving Newcomen's engine, and dutifully applied to the job the very model of methodical experimentation – alter one parameter while keeping all others constant, and see whether it helps, as indeed in several cases it did quite a good deal. Watt's original job of repairing a 1:5 scale instrument induced in him a sense of wonder, at a much more fundamental level than any attained by Smeaton, about what actually happened to the – in his unique view, outrageous – amount of fuel burned away by the machine. Luck had its part, too, in that Watt was the local handyman at the very university where the onset of a science of heat was furthest advanced in the world. Still, what makes Watt unique as an inventor in that intermediate class of what I have called 'engineers of a new kind' is that, rather than using elements of science already available, he went ahead and discovered some more himself, prior to addressing the ensuing, truly tough problem of how to make those new scientific insights in 'theoretical practice' serviceable for 'practicable practice' in its turn.

At this point, recall two key elements in my preceding analysis of the seventeenth-century gap between mathematical science and the crafts, to wit, the mathematical overshooting of the mark, and the underestimation of the world's messiness. Under the aegis of Francis Bacon, in particular, to whose message few on the Continent paid more than lipservice, British scholars had turned themselves into masters of the messy, while leaving sophisticated mathematics rather to the Continent. Newton is the great exception to this rule, of course, but note how quickly so-called Newtonianism was turned in Britain from a guide towards ever newer mathematical-experimental science into an ideology highlighting those very elements of Baconian empiricism Newton himself had felt compelled to pay lipservice to. The Continent, with culturally dominant France in the forefront from the late seventeenth century onwards, remained the locus for expertise in advanced mathematical science. Britain in its turn, with a century of somewhat haphazard, Baconian experimentation behind it, in the eighteenth century became ever more the expert in precisely the sort of relatively low-level finding and then binding together of scientific facts where the messy problems one encounters when turning 'theoretical practice' into 'practicable practice' then proved actually to be most often situated. For this is a crucial point: the very approach of tentative, certainly theory-imbued yet comparatively low-level empiricism which by the early eighteenth century had seemed to culminate in the 'extraction of sunbeams out of cucumbers' then quickly proved to be the very approach most proper for effectively to begin to bridge our gap. Add to this that, under the Baconian ideology that went under the sainted name of Newton, these efforts found some sanction in, or were at least felt to be quite compatible with, values much fostered in society at large, thus giving Britain an edge in overcoming our impediments (c) and (d), craftsmen education and decreasing social distance, and we begin to see how it is that Britain is where one of such great economic flourishing periods as lie scattered over the world's history could lead,[17] not to eventual decay as ever before, but, instead, to the invention of the very engine that was to alter for good the face of the earth.

To sum up the present point. Having been exposed in my student days not to British but to some smattering of Dutch exceptionalism, I have little ideological stake in the general affirmation-or-denial of British exceptionalism, but am rather disposed to acknowledge a specific, British exception when I think I see one. In the present case – the eighteenth century as the period when Galileo's and Bacon's aspirations began to be fulfilled; the period when our list of seventeenth-century impediments began actually to be overcome – I certainly do see one. In mathematical science (not Britain's strong point between Newton's death and Maxwell's early probings) we find a number of Continentals concerned with its application in practical practice in the tradition of Huygens and Mariotte, such as Parent, or Bélidor, but few Britons, with John Smeaton, the man to use the calculus to find out that, against intuition, overshot waterwheels greatly surpass undershot wheels in effective power, being rather the exception. But where the less mathematical sciences are concerned, Britain is where the significant inventions were made indeed. With men like Harrison, Newcomen, Smeaton, Watt, we watch the rise of engineers of a wholly new kind – men to pick up such elements of modern science as had come up in 'theoretical practice', and then fruitfully to join these to the greatest gifts of truly inventive craftsmen of the past. That is how Harrison mastered sufficient scientific theory in the domain of metal expansion, and joined it to the kind of single-minded, manual dexterity and inventive genius no scientist however capable of work with his hands could possess, so as to find the practicable solution to the problem of longitude solved long before in 'theoretical practice'. That, too, is how Newcomen did what Papin had not done – turn the idea of condensing boiling water in a cylinder and use the consequent rise of the piston for your working stroke into a practicable engine. Not counting some success stories of craft/science interaction in a few seventeenth-century scientific instruments, which really are a case apart, Newcomen's 1712 engine was the first large-scale specimen of that world-historically unprecedented phenomenon, the onset of science-based technology.

Returning now to what this means in terms of our *Economist* piece, and for its objection (2) in particular, we end up, not only by agreeing wholeheartedly *with* its author that Galileo's or Huygens' or Hooke's science was not as yet up for practical application, but also by emphasizing *against* the author that the first significant stirrings of a science-based technology are to be dated, not to Edison's times but to the early eighteenth century. What happened in the seventeenth century was in effect a learning process, which began to pay off in the eighteenth, when a new type of engineer learned to combine craft practice of the customary, trial-and-error and rules-of-thumb kind with advanced scientific insights whether picked up orally or in writing. Of that new type, Thomas Newcomen stands out as a very early, possibly the earliest representative, with men like John Harrison of chronometer fame or John Smeaton, who doubled the fire-engine's effective power, following in his footsteps. The mechanician James Watt then came up with basic scientific insights about heat and steam of his own finding, which led to so drastic a transformation of the fire-engine that henceforth it could serve to power, not just the pumping of mines which under other market conditions might also have been performed by horse- or man-power, but increasingly huge production processes. As noted, Watt in so doing brought our new type of engineer to a, once again, novel plane of achievement. Plenty of empirical material is available on the details of this entire, fascinating learning process; what seems to have been rather lacking so far is to consider it in this wider perspective. And to round off this particular issue in terms of our *Economist* piece, we can now see that its objections (1) and (2) combined, i.e., the objection arising from the gap between the Scientific Revolution and the onset of industrialization, is not really an objection to positing a close causal connection between the two events at all, but rather a stimulus towards a better understanding of how exactly the connection is to be drawn.

A CHINESE STEAM ENGINE?

Reaching the end of our own argument, it is high time to take up the issue of whether, as *The Economist* strongly implied and many an economic historian seems to affirm without much ado, at least the basics of what, in the 1760s through 1790s, James Watt wrought out of Thomas Newcomen's first piece of science-based technology might just as well have been pulled off in China. Let us grant for ease's sake that knowledge about atmospheric pressure was available in China, which surely had not been tainted with Aristotle's, in Europe so influential, conviction that air, being a light element, strives upwards. What we have then at best is a necessary, yet certainly not a sufficient condition for a Chinese steam engine or even a fire-engine. I am a great believer in the heuristic fertility of imagining alternative pathways beyond the ones actually trodden in either past or present by humankind. In my view, in history we must distinguish between three kinds of events: those that happened, those that failed to happen because they could not possibly have happened, and those that failed to happen even though realistically speaking they might well have happened. Whereas I think it can be argued that some Galileo-like achievement by the end of the Golden Age of Islamic science is among the alternative pathways that might in principle have been taken, I shall now set forth with utmost succinctness why I think a Chinese steam engine belongs rather to my second category of historical events – the (given the state of affairs at that particular time and place) inherently impossible ones. Given the overall tenor of nature-knowledge in China, I cannot possibly imagine what such an alternative pathway might have looked like. In particular, I fail to see how such a piece of machinery might ever have been constructed in the absence of the kind of counter-intuitive knowledge of the void, and in the absence of the defiance of direct perceptual knowledge of nature that goes with it, which only incipient modern science in Europe began to provide. What is so special about the fire-engine and, *a fortiori*, the steam engine, is not that in how it works usage is being made of the action of certain scientific principles. Such is also the case for a celt or a bellows. The discriminating question is *whether the usage was conscious* and, if so, *whether that consciousness was necessarily required*. Take the case of the mixture stops in late medieval organs I have mentioned before. Here it was quite possible to put to marvellously productive usage scientific principles unknown as such, with the effect being achieved by arranging and subtly cutting up certain kinds of organ pipes in a purely empirical, trial-and-error manner. So, the true question to ask about a possible, alternative pathway for the steam engine is whether it might have been pulled off with condensation of steam in a vessel and atmospheric pressure being exerted upon the void thus created, without knowing about it. Might have happened with the fire-engine what happened with those mixture stops, i.e., craft construction upon an unknown scientific foundation? My mind boggles at the notion that it might ever have occurred to anyone anywhere to put together, just gropingly, a boiler, a cylinder, a piston and an assembly fit to ensure regular condensation-cycles in a space regularly drawn void. Maybe it is just my mind that boggles, but at the very least I think that the burden of proof in the present case rests upon those who fancy that there is no problem here at all, and that the West owes the prime power mover behind significant stages of its industrialization process to sheer chance. None other than Joseph Needham, in the 'Newcomen Centenary Lecture' he delivered in 1963 for the Newcomen Society under the title 'The Pre-Natal History of the Steam-Engine', took the inconceivability of such a hypothetical occurrence wholly for granted, which is the more significant as Needham rarely if ever forwent a chance to raise a claim for Chinese priority and general excellence in science and technology. In this particular paper he argued no more than that James Watt, both for making his steam engine double-acting and for converting its to-and-fro motion into rotary motion, may well have made unconscious usage of certain techniques devised first in premodern China and then transmitted to Europe along a hypothetical yet (in Needham's reconstruction) plausible route. So that, in his peroration, Needham concluded that:

no single man was 'the father of the steam-engine'; no single civilisation either ... Yet no one comes nearer to deserving the title than Thomas Newcomen. In the light of the foregoing analysis he stands out as a typical figure of that modern science and technology which grew up in Europe only, while his successors, great as they were, drew upon older Asian inventions more than has hitherto been recognised.[18]

MODERN SCIENCE AND THE WESTERN ORIGINS OF THE MODERN WORLD

If we now survey one final time the three highly pertinent objections raised in our *Economist* piece against the idea (which it grants at the outset to be so plausible at first sight) that modern science and technology were highly instrumental in bringing forth our modern world, we conclude that that idea, for all those well-chosen objections against it, keeps expressing a profound, albeit far too often underestimated or ignored, historical truth. Yes, science and technology and the relation between them are subtler than is still often being thought; yes, the pioneering period of our modern science is not when our modern, science-based technology came into being. But no, the advent of that modern, science-based technology did not have to await Edison. Yes, there are craft elements in the Industrial Revolution that could do well without modern science or any science at all (notably the first machines for mass spinning and weaving), though even there modern science quickly became instrumental, too (e.g., in providing the chemicals indispensable for mass bleaching and dyeing). But no, the Industrial Revolution as the event *par excellence* to usher in our modern world cannot even be thought of if a science-based technology, and, hence, the kind of science on which to base such a technology in the first place, had not come into being before. And no, there is no possible way for products of science-based technology to have come into being through any other route than that of modern science as a necessary (though not, indeed, as a by itself sufficient) condition.

Which conclusions at once raise two further questions of major proportions. Here is not the place to seek to resolve them, but at least they ought here to be signalled.

First major question: What about China? Even if it be granted that the steam engine requires modern science as an indispensable precondition, a good part of my entire argument would fall flat after all if it could successfully be argued that our modern science might just as well have emerged in China or in any of the other great premodern civilizations, with its actual emergence in seventeenth-century Europe being just a chance event as many an economic historian is pleased to maintain or at least to take for granted. I cannot of course make that point here, but only assure the reader that, for whatever it is worth, I have made it elsewhere.[19]

Second, and final, major question: What turned Watt's steam engine from a small-scale invention into a real-life machine and, from there, into a highly productive one? In the above we have seen, however crudely, how Europe by mid-eighteenth century had arrived at the brink of a viable technology erected on occasion upon a previously unheard-of, scientific basis. For all this to usher in the onset of industrialization, however, something more was of course indispensably required – the economic opportunity that actually presented itself in late eighteenth-century Britain in the guise of a readiness to *invest* in such technological novelties as presented themselves. And in many though not all respects an independent chain of argument can be (and actually has often been) set up to explain the emergence of that economic opportunity leading to those investments and the productive success thereof. The historical question that continues to baffle me more than any other, then, is this: How, next, to explain this *confluence* of two not fully yet by and large independent streams of historical events? Why should the retrospectively required mode of science *and* the retrospectively required need-*cum*-readiness for large-scale investment be there at just the same place at just the same time? There are answers to that question of questions, too,[20] but not ones that still fit in the frame of the issue here addressed of the onset of a science-based technology neither with Edison nor in China or elsewhere, but in eighteenth-century Britain as the outcome of a major learning process instigated by the Europe-wide Scientific Revolution of the seventeenth century.

Notes and References

1. *The Economist*, 31 December 1999 (vol. 353, no. 8151), 10–12. The cover offers as dates 'January 1st 1000–December 31st 1999', and carries the following slogans: 'Millennium special edition' and 'Reporting on a thousand years'.

2. H.F. Cohen, *The Scientific Revolution. A Historiographical Inquiry* (Chicago, 1994), 13.

3. Problems (1)–(4) form the subject of my near-completed book provisionally entitled 'How Modern Science Came Into the World. Three Revolutionary Transformations and the Dynamics Behind Them'. I have treated problem (5) in an article in Dutch, 'Het ontstaan van onze moderne wereld: wat natuurwetenschap en techniek ermee van doen hadden', *Theoretische Geschiedenis* 25, 4 (1998), 322–49.

4. The explanation of the action of Newcomen's fire engine that follows here owes most to D.S.L. Cardwell, *Turning Points in Western Technology* (New York, 1972), 66–72 (also *idem, The Fontana History of Technology* (London, 1994), p. 121–5)). Much detail about the snap action of the valves and cocks and about a whole range of later varieties of the engine is provided in L.T.C. Rolt and J.S. Allen, *The Steam Engine of Thomas Newcomen* (Hartington, 1977).

5. Engraving by H. Beighton, 1717, reproduced from R.J. Law, 'The Steam Engine. A Brief History of the Reciprocating Engine' (London: Science Museum Monograph, 1965), 9.

6. Reproduced from J. Needham, *Clerks and Craftsmen in China and the West. Lectures and Addresses on the History of Science and Technology* (Cambridge 1970), 211.

7. Reproduced from R.J. Law, 'The Steam Engine. A Brief History of the Reciprocating Engine'. (London: Science Museum Monograph, 1965), 9.

8. The indispensability of the Scientific Revolution for at least some key events and developments that went into the making of the Industrial Revolution has been defended previously in rather different ways by A.E. Musson and Eric Robinson, *Science and Technology in the Industrial Revolution* (Manchester, 1969), and by Margaret C. Jacob, *Scientific Culture and the Making of the Industrial West* (Oxford, 1997).

9. The expression alludes to A.R. Hall's paper 'The Scholar and the Craftsman in the Scientific Revolution', in M. Clagett (ed.), *Critical Problems in the History of Science. Proceedings of the Institute for the History of Science at the University of Wisconsin, September 1–11, 1957* (Madison,

1959), 3–23. Together with many another contribution to the debate about science and the crafts during the Scientific Revolution, I have discussed it in my *The Scientific Revolution. A Historiographical Inquiry* (Chicago, 1994), section 3.4.3.

10. The section that follows is taken almost fully from my book-in-progress mentioned in note 3.

11. L. Reti made this point in his contribution to the article 'Leonardo da Vinci' in the *Dictionary of Scientific Biography*, vol. 8, 206–14.

12. D.S.L. Cardwell, *Turning Points in Western Technology. A Study of Technology, Science and History* (New York, 1972), 42–3.

13. *Ibid.*, 38–9, and *idem*, *The Fontana History of Technology* (London, 1994), 87.

14. Christiaan Huygens, *Oeuvres Complètes* 9, 272–91; p. 289 (report, in Dutch, to the Dutch East India Company of 1688).

15. R.S. Westfall, 'Robert Hooke, Mechanical Technology, and Scientific Investigation', in J.G. Burke (ed.), *The Uses of Science in the Age of Newton* (Berkeley, 1983), 85–110.

16. D.S.L. Cardwell, *Turning Points in Western Technology* (New York, 1972), 83–9; also enlightening on the 'science' portions in Watt's accomplishment is Arthur L. Donovan, 'Toward a Social History of Technological Ideas: Joseph Black, James Watt, and the Separate Condenser', in G. Bugliarello and D.B. Doner (eds.), *The History and Philosophy of Technology* (Urbana/Chicago, 1979), 19–30.

17. This phrase alludes to the core thesis of J. Goldstone, 'Efflorescences and Economic Growth in World History: Reframing the "Rise of the West" and the British Industrial Revolution', *Journal of World History*, October 2002.

18. Joseph Needham, *Clerks and Craftsmen in China and the West. Lectures and Addresses on the History of Science and Technology* (Cambridge, 1970), 136–202; quotation on 202.

19. In my book-in-progress mentioned in note 3.

20. Elements of such answers are to be found in work by Margaret C. Jacob and in J. Uglow, *The Lunar Men. The Friends Who Made the Future* (London, 2002).

The Emergence of Science-based Technology. Comments on Floris Cohen's Paper

ALESSANDRO NUVOLARI

In his paper Floris Cohen raises a very important question, to which historians have so far devoted but scant attention, namely, why the emergence of science-based technology did not follow immediately from the Scientific Revolution of the seventeenth century ? In this respect, it is frequently held that traditional arts and crafts (based on trial-and-error and slow accumulation of rules of thumb) persisted well into the nineteenth century as the major source of technical advances.

It is interesting to note that the existence of this interlude between the 'beginnings' of modern science and the acceleration of economic growth brought about by the Industrial Revolution, has made it difficult for historians to properly assess the role played by the emergence of 'science-based' technology in the early phases of industrialization. As a matter of fact, economic historians have frequently stressed the empirical rather than scientific nature of most of the technological advances of the Industrial Revolution. In this view, modern science begun to contribute substantially to economic growth only from the late nineteenth century with the creation of industrial research and development laboratories.[1]

In my judgement, this view is essentially flawed, being based on an essentially incorrect understanding of the emergence of science-related technology and of its critical contribution, by means of a complex set of ramifications, to the process of modern economic growth. In his paper Cohen examines in detail the process of 'rapprochement' between modern science and technology that took place during the eighteenth century. According to Cohen, a salient feature of this process was the emergence of a new type of craftsman, capable of utilizing the insights of 'new' science in his practice. In this sense, Newcomen, Smeaton and Watt (although their approaches are characterized by interesting differences) represent archetypical figures marking the rise, in the course of the eighteenth century, of a new type of (scientific oriented) craftsmanship.

Unfortunately, we have very little information on the actual origins of Newcomen's invention. It is reasonable to presume that Newcomen and his partner John Calley had some knowledge of the Savery engine.[2] In turn, it is quite clear that Savery's invention was consciously making use to contemporary scientific investigations on atmospheric pressure. Indeed, one could say that Savery and Newcomen concluded a long process of gestation, making the idea of harnessing atmospheric pressure (which before had remained confined to the scientist's laboratory) susceptible to practical application.

Contrary to Savery, Newcomen was capable of devising a design which was fully within the limited engineering capabilities of the time. The historical significance of Newcomen's invention has been tersely summarized by Donald Cardwell:

> [The] virtues [of the Newcomen engine] were evident and out-standing: it could be fabricated by local craftsmen without requiring any techniques other than those to which they were well accustomed; only the cylinder and one or two other components might have to come from 'specialist' manufacturers – like the famous ironworks at Coalbrookdale – and even the cylinder made little demand on the skills of the time beyond those already acquired for the noble arts of casting and boring cannons. The nature of the beam engine was such that its components did not have to be very accurately aligned. It was simple to operate, no great skill, and certainly no science, being needed by the engineman; and in a country wholly without technical education, this was no small thing.[3]

Given these merits, the rapid and geographically wide diffusion of the Newcomen engine in Britain does not come as a surprise.[4] What is worth remarking, in this respect, is that the rapid diffusion of the Newcomen engine gives an indication of the rich endowment of mechanical skills existing in Britain. Of course, a large number of these mechanics were not engineers of the calibre of Newcomen, Smeaton and Watt. However, the available evidence suggests that trial-and-error and rules of thumb had begun to be increasingly coupled to a more scientific approach. William Fairbairn described the world of the millwright around 1750 in a passage that is particularly illuminating:

> The millwright of former days was to a great extent the sole representative of mechanical art, and was looked upon as the authority on all applications of wind and water, under whatever conditions they were used, as motive power for the purposes of manufacture. He was the engineer of the district in which he lived, a kind of Jack-of-all-trades, who could with equal facility work at the lathe, the anvil, or the carpenter's bench. ... Thus the millwright of the last century was an itinerant engineer and mechanic of high reputation. In the practice of his profession he had mainly to depend on his own resources. Generally, he was a fair arithmetician, knew something of geometry, levelling and mensuration and in some cases

possessed a very competent knowledge of practical mathematics. He could calculate the velocities, strength and power of machines; could draw in plan and in section and could construct buildings, conduits or watercourse, in all the forms and conditions required in his professional practice. ... Such was the character and condition of the men who designed and carried out most of the mechanical work of this country to the middle and end of the last century.[5]

As Joel Mokyr has pointed out, the abundance of mechanical skills was one of the main determinants of Britain's technological advantage in the early phases of European industrialization.[6] Following Cohen, I would qualify Mokyr's contention by saying that Britain's advantage rested on her rich endowment of these *new types* of artisan and mechanic.

I would also like to suggest that the emergence of modern scientific oriented craftsmanship was based on two rather distinct sources.[7] The first one is the traditional world of millwrights and iron makers, working mainly with wood and wrought iron. Newcomen (an ironmonger) and his partner Calley clearly belong to this tradition. As we have seen, one of the main merits of Newcomen's invention was its quality of being fully within the limit of a typical millwright's mechanical capabilities. The second source is the world of clock and scientific-instrument makers. These were artisans who adopted high levels of accuracy in the components of the artefacts they produced. The main material in those 'trades' was brass, a metal which could be very precisely processed. It is interesting to notice that inventions emerging from this tradition frequently call for a *stretching* of existing engineering capabilities. For example, Watt's separate condenser (Watt was a scientific-instrument maker at the University of Glasgow) required a high degree of accuracy in the boring of cylinders. As is well known, Watt's invention had to be coupled with the concomitant invention of the cylinder-boring machine by John Wilkinson (an inventor coming, instead, from the iron makers' tradition). Thus, we have a complex interplay between the application of scientific knowledge to the problems of industrial production, the improvement of traditional mechanical skills and instruments, and the diffusion and application of new technologies. From the second half of the eighteenth century, in many branches of manufacturing, this interaction would mature into a continuous self-reinforcing process.

To conclude, Newcomen's invention, as Cohen suggests in his paper, shows that from the early eighteenth century, 'new' arts and crafts capable of producing science-based technologies had begun to emerge. The origins of this process in the course of the seventeenth century are still rather obscure. It is another merit of Cohen's paper to flag a promising and still largely unexplored research agenda.

Notes and References

1. This passage from McCloskey is a good example: 'A powerful myth of moderns is that Science Did It, making us rich. The finding of Not is ... relatively recent ... . Few parts of the economy used much in the way of applied science in other than ornamental fashion

until well into the twentieth century. In short, most of the industrial change was accomplished with no help from academic science' D. McCloskey, '1780–1860: a survey', in R. Floud and D. McCloskey (eds), *The Economic History of Britain since 1700* (Cambridge 1994), 266.

2. On the possible contacts and exchanges between Savery and Newcomen, see L.T.C. Rolt and J.S. Allen, *The Steam Engine of Thomas Newcomen* (Hartington, 1977).

3. D.S.L. Cardwell, *Steam Power in the Eighteenth Century. A Case Study in the Application of Science* (London, 1963), 33.

4. J.W. Kanefsky and J. Robey, 'Steam engines in 18th-century Britain: a quantitative assessment', *Technology and Culture*, 21, 161–86.

5. W. Fairbairn, *The Life of Sir William Fairbairn* (London, 1877), 26–7.

6. J. Mokyr, 'The new economic history and the Industrial Revolution' in J. Mokyr (ed.), *The British Industrial Revolution: An Economic Perspective* (Westview, 1993). It is also worth remarking that the 'organization of science' in Britain during the eighteenth century was characterized by what may be called a process of 'democratization', in the sense of increasing openness to different social groups (including artisans and mechanics). On the peculiar, 'non elitist' character of eighteenth-century British science, in comparison to other European countries, see I. Inkster, *Science and Technology in History* (London, 1991), ch. 2.

7. A similar point has been suggested by C.B. Cragg, 'The evolution of the steam engine', in K. Hahlweg and C.A. Hooker (eds.), *Issues in Evolutionary Epistemology* (Albany, 1989).

The Formation of Knowledge Concerning Atmospheric Pressure and Steam Power in Europe from Aleotti (1589) to Papin (1690)

GRAHAM HOLLISTER-SHORT

When Federico Commandino's Latin translation of Heron of Alexandria's *Pneumatica* appeared in 1575 it may well have introduced Renaissance scholars to novel material but it is abundantly clear, at the level of know-how, that Giovanni Aleotti's translation of Commandino's work of 1589 into Italian can have conveyed little new information for the artisans for whom Heron in the vernacular was presumably intended.[1] The fact of the matter was that the working procedures (that is, the sophisticated constructional know-how) which lay behind Heron's devices, and indeed the devices themselves, were well understood long before 1589 (Figure 1).

Heron's *Pneumatica*, a work composed about AD 50, had been translated two or even three times in the twelfth and thirteenth centuries in Sicily, although none of these translations seems to have survived down to the sixteenth century.[2] These versions were one channel through which Heron's work became known to the Latin West. Another channel may very probably have been furnished by intimate contacts with the Islamic world where the entire Greek corpus of works on machines had been translated into Arabic, assiduously cultivated and elaborated throughout the entire period of the Middle Ages.[3] At all events, the principal effects achievable by means of rarefied, compressed and heated air can all be shown to have been known, both theoretically and in practice, in the West well before 1500. Sufflators (aeolipiles) for providing blast for furnaces and alembics, often made in anthropomorphic forms (rather like Giovanni Branca's blower figured in his *Le Machine* of 1629) were known in the Middle Ages, siphons likewise, and force pumps, which had probably never gone out of use since Roman times. Force pumps had indeed been

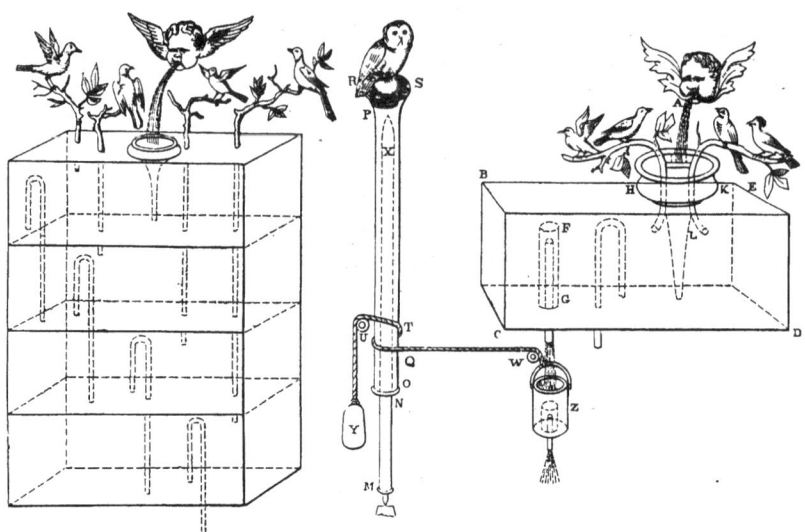

Figure 1   Bennet Woodcroft (ed.), *The Pneumatics of Hero of Alexandria* (London, 1851), p. 31 (device no. 15) and p. 67 (device no. 44). The singing birds at Tivoli were supplied by a compartmented trompe (but of more sophisticated form) with siphons similar to device 44, with the addition of the owl device (but modified to produce horizontal travel) of device no. 15.

supplemented by a new form, the suction lift pump, which by about 1425 appears in virtually fully-fledged form in the work of Mariano Taccola.[4] As for the use of falling water to entrain air into closed vessels (trompes) by means of aspirators, that too appears to have been known in the Middle Ages. The metal tree with singing birds, fashioned by Guillaume Boucher of Paris for the Great Khan in Karakorum and seen by de Rubruquis some time in 1346, suggests as much, while the use of the trompe (as well as much else) certainly explains a passage in Leonardo's notebooks composed *c.*1500. In his plans for a garden, probably drafted while he was in the service of Cesare Borgia in 1502–3, he wrote that he knew how to make all manner of mills for producing music and how to create the means for producing the songs of birds.[5] As for the theory upon which all these devices were thought to depend, given the unchallenged sway of the Aristotelian postulate of the plenum, Heron's ideas had to serve as explanation. This was that while no continuous vacuum could exist in nature, matter itself was particulate, the particles separated by micro-vacua, something which permitted air to be compressed or rarefied. In this way the 78 various effects that he described and illustrated in the *Pneumatica* could be explained after a fashion.

By the 1570s it has been suggested that trompes were widely known in Italy and were used for providing the blast for metalworkers' forges, but given the present state of knowledge this seems hardly likely. The importation of French experts to construct the works at Tivoli certainly suggests otherwise. The evidence as to the signification of the word 'trompe' carefully collected by Enzo Baraldi for the mountainous area around Brescia and Bergamo points very strongly to the fact that the word only began to mean a blowing device at the beginning of the seventeenth century. The earliest date found by Baraldi for its use in smithies is 1627.[6] As for the evidence offered by Jean Errard's figure of such a device in his *Premier livre des instruments mathematiques mechaniques* of 1584, it seems perhaps that knowledge of the device was equally scanty in north-east France at much the same time. The caption to his engraving defines the word 'trompe' adequately enough: 'a device by means of which smiths are able, with cold water, to excite fire with wind in place of bellows', but either through ignorance or deliberate obfuscation (a not untypical strategem used by authors of machine books), the engraving displays a woeful lack of real knowledge of the device (Figure 2). There is no doubt, however, that for metallurgical, as indeed for thaumaturgical, purposes a prime consideration was an unfluctuating, sustained blast so that once the technique was understood there would be every incentive to employ it. The blast produced by a well-constructed trompe would have been remarkably steady, with none of the fluctuations forge masters experienced with the better-known use of pairs of single-acting cuneate bellows working alternately. The nine-teenth-century metallurgist, John Percy, quoting Georges Richard, has this to say of what was then merely a vestigial reminder of former practice: '... no machine produces a blast of such constant tension as the trompe, the mercury appearing as though it were congealed in the gauge.'[7]

But to return to Aleotti; had he nothing new to add to the Heronic corpus in 1589? In one respect, yes. Annexed to his translation of Heron are four theorems in which he sought to display his own virtuosity. The first three are really re-combinations of elements present in Heron. The fourth theorem, however, reveals something novel. This involved the exploitation of the refrigerative power of compressed air expanding on its release through narrow-gauge tubes. Accordingly, we find Aleotti describing how a room could be kept deliciously cool during the burning days of summer by this means.[8]

In explaining how these devices worked Aleotti bases himself, not without some confusion, on Heron's ideas regarding the composition of matter. Nature did not permit a continuous vacuum to exist but Aleotti was somewhat ambivalent about following Heron further in admitting the compressibility of air and its rarefaction. In a manuscript composed about 1615, and examined by Alex Keller some years ago, Aleotti pointed to the phenomena observable when a ramrod was used to compress the charge in a cannon. It was difficult to force down the ramrod when the touch-hole was stopped up because air could not be compressed, and difficult to extract when the ramrod was in the barrel and the touch-hole was then stopped up because the extraction of the ramrod would have begun to create a vacuum. Since Aleotti was unwilling to entertain either the idea of the vacuum or of micro-vacua existing in nature, this meant that Aleotti could not explain why the devices he himself had created were able to work.[9]

Figure 2 Jean Errard, *Le premier livre des instruments mathematiques mechaniques* (Nancy, 1584), fig. 31. The caption runs: 'Artificium, quo fabri aqua calida spiritu ignem loco follium excitare possunt'.

But let us look more closely at the experiences with the cannon described by Aleotti about 1615. As for the non-compressibility of air, when the touch-hole of a cannon was stopped up it was extremely difficult to attempt to force a ramrod down the barrel, and if it *were* thrust down as far as it would go and then released, the air would burst out and expel the rod furiously. Conversely, in respect of rarefaction, if the ramrod had been fully inserted while the touch-hole was open, and this was then stopped up, it became equally hard to draw the ramrod up. If the rod were then released by those pulling on it, it would of itself retract forcibly. These experiences I shall return to presently. But why was Aleotti so exercised about Heron at all? One is very tempted to look at what Cardinal Ippolito II d'Este of Ferrara had set his hand to at Tivoli, some 20 miles north of Rome, once he had been appointed governor of the place in 1550. On the precipitous slope falling away from the front elevation of the palace at the Villa d'Este he had given Pirro Ligorio a free hand to create a water garden of Heronic delights, and this on the largest scale. Its hydro-mechanical delights, I believe, stimulated both Commandino and Aleotti to publish, as we might say, the books of the garden, or books which at least made available the technological antecedents of what was so spectacularly on view at Tivoli. The hydro-mechanical effects, driven by water derived from the Teverone, were the work of two French artisans, Luc Leclerc (d. 1572) and Claude Venard, his nephew. There are a number of accounts of the garden but that of Nicolas Audebert, who was a guest at Tivoli some time either late in 1576 or early in 1577, has the extra importance of being a fully informed eye-witness account. In this respect Audebert was indeed specially privileged. As a fellow countryman of Venard's, then in charge of the devices following Leclerc's death, he was admitted to the secret parts of the machinery not normally shown to visitors (Figure 3). Audebert was even shown the great arcanum of the garden – how the mechanical music was produced – and was allowed to climb down into the chamber of the trompe in which was stored the compressed air for the hydraulic organ. Another trompe elsewhere set a multitude of mechanical birds singing, each producing its natural range of notes.

At yet another of the many fountains a further trompe was used to produce a range of sound effects that mimicked musket fire and the discharge of cannon. These were produced by the violent explusion of air as large volumes of water were admitted into pipework concealed in the figures of dragons and marine creatures.[10] I mention the work of Leclerc and Venard because it shows a living, artisanal knowledge not only of how Heronic devices might be deployed some years before the earliest translations of Heron became available, that is, from 1575 onwards, but also how those devices might be married to a mastery of 'tonotechnie'.[11]

And so things continued with della Porta's *Pneumaticorum Libri Tres* of 1601, with Salomon de Caus' work of 1615 and with that of Giovanni Branca in 1629.[12] The great fault line separating them all from the later work of Otto von Guericke, Huygens and Papin was the seismic shift

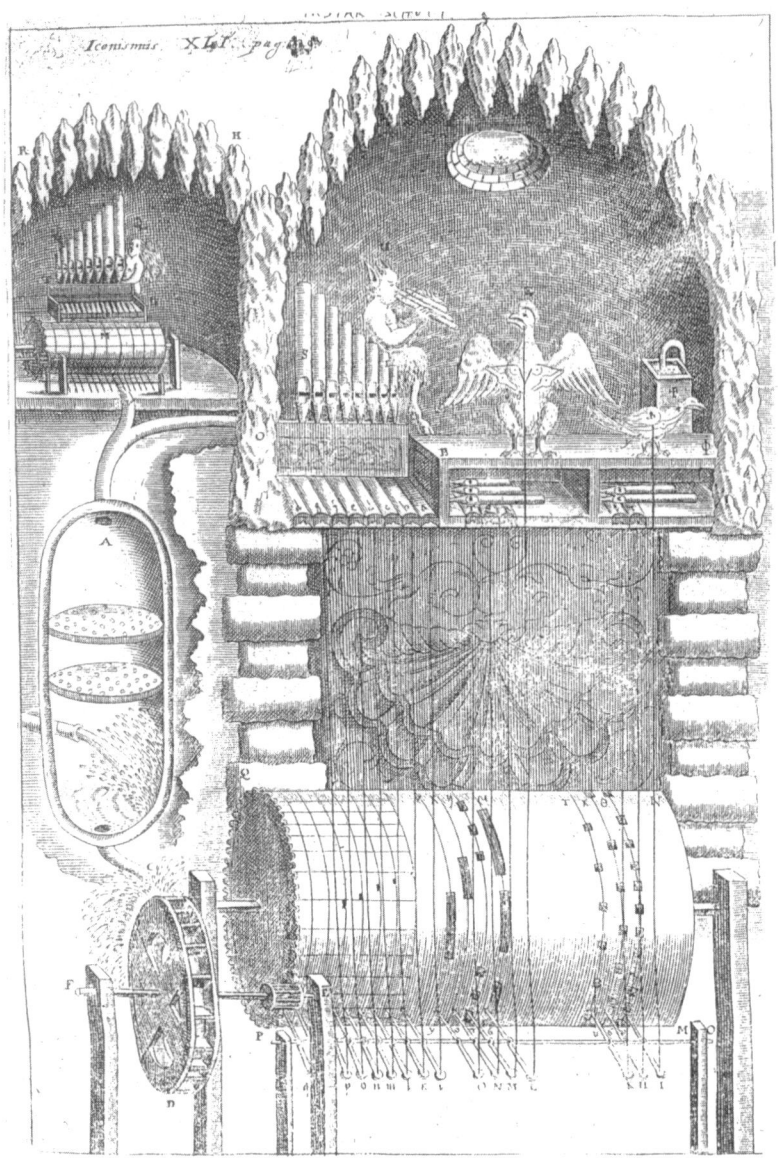

Figure 3 Gaspar Schott, *Mechanica Hydraulico-Pneumatica* (Würzburg, 1657), p. 424, Iconismus XLI.

induced by Evangelista Torricelli's experiments of 1644 in establishing the role of atmospheric pressure and the weight of the air in explaining the action of suction pumps and of siphons, as also the manner in which the so-called Torricellian vacuum might be produced. All of this, and the crucial experiment which established it, were conveyed by Torricelli in his letter to Michelangelo Ricci of 11 June 1644.[13] It is in this letter that he uses the memorable phrase: 'Noi viviamo nel fundo d'un pelago d'aria elementare' – 'We live at the bottom of an ocean of elemental air', an ocean which he estimated was some 15 miles deep. By various channels, intelligence of Torricelli's findings was soon diffused to every part of learned Europe, there to be accepted by all save those committed to upholding the teaching of the Roman Church.

By some route not known to us – or so it would seem, unless we are willing to suppose that he pursued a completely independent line of enquiry – Otto von Guericke learnt of Torricelli's results and began experiments using an air pump of his own devising to exhaust the air from specially constructed spheres (the so-called Magdeburg spheres).[14] Conventionally he is held to have begun this work about 1650. By 1654 von Guericke's experimentation had led to his development of a piston and cylinder device. This enabled him to demonstrate quantitatively, although in a somewhat rough-and-ready way, how much weight the atmosphere exerted on the piston when a vacuum was created in the cylinder in which it worked.[15] When, in 1672, von Guericke published details of further work along these lines, he was able to report even more impressive results. The engraving which accompanied his description figures what might perhaps be described as a proto-atmospheric engine. First one attached an evacuated sphere (fitted with a tap) to a cylinder and piston device. Next von Guericke was able to demonstrate that on turning the tap and opening communication between the sphere and the cylinder, the drop in internal pressure in the cylinder, as the air within it rushed into the empty sphere, caused the piston at the top of the cylinder to be pushed down by atmospheric pressure. By this means a weight of 2680 pounds (*c.* 1200 kg) could be raised (Figure 4). Some years before this, however, Gaspar Schott, in the course of writing his *Mechanica Hydraulico-Pneumatica* of 1657, had entered into correspondence with von Guericke. He subsequently published the series of questions and answers that had passed between them as an appendix to the *Mechanica*. ... The sequence conveys a splendid sense of immediacy. In the course of these exchanges von Guericke, in a letter of 22 July 1656, describes how after prolonged pumping, 'post longam agitationem', it was still possible to move the piston, 'adhuc posse agitari'. However, if the piston handle of the pump were released by the two burly men who had pulled it up, it was propelled back, 'retrudit intus', into the body of the pump cylinder with some violence. Von Guericke's account of all this, as reported by Schott, is of the greatest interest. If the cylinder so evacuated was one ell (24 inches) in diameter, it needed a pull of over 1200 pounds, exerted by the combined effort of six labourers, to pull the piston back to the top of it.[16]

I should like here to recall Aleotti's explanation concerning the ramrod and the touch-hole of the cannon mentioned above. When the touch-hole was stopped up, it was incredibly difficult to pull up the ramrod, and if the ramrod were released, it was pulled back forcibly (as Aleotti might have put it) into the barrel of the gun. Now let us imagine Aleotti and von Guericke both

Figure 4   Otto von Guericke, *Experimenta Nova, ut vocantur, Magdeburgica de Vacuo Spatio* (Amsterdam, 1672), Bk 3, ch. 27, Iconismus XIV.

watching the two men pumping the air out of the cylinder. On the face of it both see the same thing when the pump handle is released. Each of them, however, has a totally different explanation as to what is causing the pump to perform a stroke of itself: one sees it as nature resisting the creation of a vacuum, the other as the weight of the atmosphere acting against reduced pressure. To make the change from denial of the possibility of a vacuum to recognition that a vacuum could be created at will (as in Torricelli's mercury-filled tube when inverted) was to experience a gestalt shift of enormous significance. To accept the new situation was a necessary precondition for the work of those who followed Torricelli. The mental landscape of seventeenth-century Europe was riven by such switches. In a similar way to this imaginary confrontation of Aleotti (it could equally well have been Schott) and von Guericke, Russell Hanson in his *Patterns of Discovery* invites one to picture Tycho Brahe and Kepler on a hilltop at dawn facing east. When the sun appears, do they both see the same thing? Do they both see the sun rising or does one of them see the earth tilting?[17]

The publication of von Guericke's *Experimenta Nova, ut vocantur, Magdeburgica de Vacuo Spatio* in 1672 (in which a much more sophisticated version of the weight-lifting rig of 1654 takes its place) seems likely to have been the event which set the stage for Christiaan Huygens to think about how to exploit the vacuum in a less labour-intensive way than von Guericke's. His invention in early 1673 of a gunpowder engine was designed to do just that.[18] Instead of having two burly men labour with an air pump for many hours to create a reasonable vacuum, Huygens dreamt of creating a rapid sequence of vacuums by exploding very small quantities of gunpowder at a brisk tempo. This would permit the creation of a machine with a high stroke rate. The *modus operandi* of Huygens' machine is easily explained. A piston is held at the top of a tube. A small charge of gunpowder is ignited at the base of the tube, the expanding gases evacuate themselves through side valves. The resulting drop in pressure in the tube causes atmospheric pressure to force the piston down into the cylinder. This was the power stroke in a single-acting machine, the piston being returned to its original position by the counterweight of the apparatus against which it had been acting (Figure 5). Roughly speaking, after the gases of the explosion had blown off, pressure in a cylinder having a diameter of some 12 inches fell to only about half an atmosphere. Despite repeated efforts to overcome this problem – Huygens even resorting to the use of a fusée – in the end only disappointingly few power strokes could be achieved before the residues remaining in the cylinder were compressed up to a value equal to atmospheric pressure.

Huygens' assistant, Denis Papin, worked on this project equally without success. By 1690, however, he had come to realize that steam might be used as an expansive medium in place of the gas released by the ignition of gunpowder.[19] Unfortunately, or so it seems to me, the plan on which he proceeded to exploit this idea did not derive from von Guericke's weight-lifting rig as I suppose Christiaan Huygens' gunpowder engine to have done, or, for that matter, from his old master's apparatus. In Papin's device steam is not used as gunpowder had been – that is, passively and merely as a way of creating a vacuum – but is exploited expansively to push the piston from the bottom to the top of the cylinder (Figure 6). At the top of its travel the piston is arrested by a catch until condensation has occurred, whereupon the catch is released and atmospheric pressure then permits a further power stroke to follow. Papin was thinking in terms of a double-acting engine: fitting a toothed piston rod of the device engaging with a toothed wheel would be the means of achieving power strokes on both excursions of the piston. This was not, however, to be the path to a successful engine. In this respect Thomas Newcomen's single-acting machine, using steam only as a means to an end, follows the devices of von Guericke and Huygens in its basic mode of operation, thereby placing a question mark over just how much Newcomen might have derived from Papin. But *non uno itinere*. How, for instance, should one attempt to fit into this story the sketch of a steam engine in Roger North's MS: *R. North Pictures, Engines & Inventions*?[20]

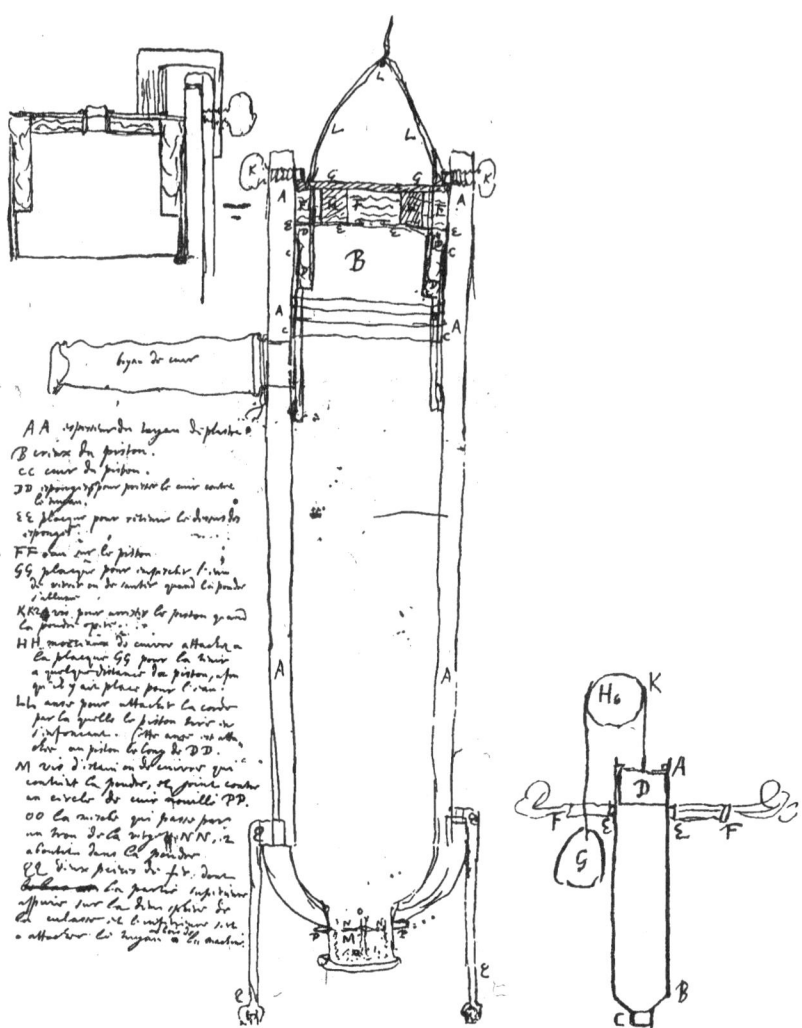

Figure 5 Christiaan Huygens, (i) drawing of 10 February 1673, *Oeuvres Complètes* (The Hague, 1950), Vol. 22, p. 243; and (ii) the simplified sketch of 22 September, 1673 sent to his brother, *Oeuvres Complètes* (The Hague, 1897), Vol. 7, p. 357.

But why had Papin strayed from what, with hindsight, was to be the correct path (initially at least) towards the exploitation of steam? The answer lies in what I should like to call a kind of thrift economy practised by Papin. Nothing Papin saw seems ever to have been discarded. On 9 September 1663 Robert Hooke presented to the Fellows of the Royal Society a prototype gunpowder engine for measuring the work done when a given quantity of gunpowder was exploded (Figure 7). Hooke achieved this by means of a weighted lever moved by a toothed ratchet.[21] The

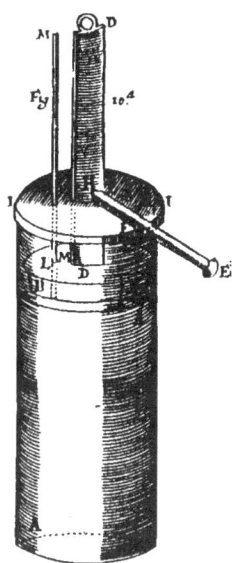

Figure 6 Denis Papin, 'Nova methodus ad vires Motrices validissimas levi pretio comparandas', *Acta Eruditorum*, 1690, p. 410.

weighted lever reappears as part of a 'safety valve' in Papin's digester, or pressure cooker, of 1679, likewise displayed before the Royal Society. But the 'safety valve' itself (a small piston in the cover of the digester) pushing against a weighted lever à la Hooke is in reality still only a means of permitting the pressure of the steam within the digester to be measured. It was not, in any real sense of the word, a safety valve at all. Papin's elaborate experiments with the *marmite* revealed to him the quantity of

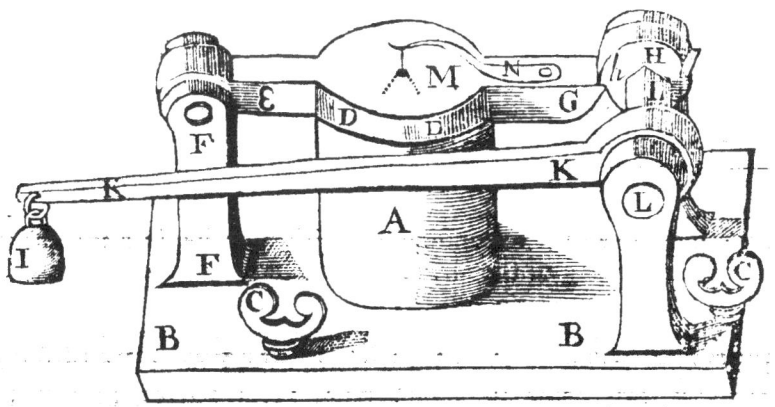

Figure 7 Thomas Birch, *The History of the Royal Society of London* (London, 1756), Vol. 1, p. 302.

work that even a small piston, only two-fifths of an inch in diameter in the case of the digester, could perform when pushed upwards by steam at high pressure. My suggestion is that Papin's 'steam engine' of 1690 is simply a scaled-up version of the 'safety valve' idea of 1679. What lends substance to this idea, if it is accepted as a working hypothesis, is that one can then see that Papin, in making the conceptual leap from digester to steam engine, was unwittingly carrying across a skeuomorphic feature, that is, an unnecessary element, into the new device. For the digester to function, steam had necessarily to push the piston/safety valve upwards but this was in no sense a necessary feature of an atmospheric steam engine. The piston in the model of 1690 has a diameter of only two inches, so that scaling up had not proceeded very far. What was nevertheless abundantly clear to Papin was the prodigious power-weight ratio (something like 240: 1) of his 'engine'. The device, however, was shaped by the form of Hooke's proto-engine rather than by Huygens' gunpowder engine (the latter based, as I have suggested above, on the model of von Guericke's device). This might seem to explain how Papin had wandered off the track that was to lead in the course of the next 20 years to Thomas Newcomen's atmospheric engine.

Notes and References

1. F. Commandino, *Heronis Alexandrini spiritalium liber, a Federico Commandino Urbinate, ex Graeco nuper in Latinum conversus* (Urbino, 1575); G.B. Aleotti, *Gli artificiosi moti spiritale di Herrone ... Aggiuntovi dal medesimo quattro theoremi non men belli et curiosi de'gli altri* (Ferrara, 1589).

2. M. Boas, 'Hero's Pneumatica. A study of its transmission and influence', *Isis*, 1949, 40: 38–48, especially 40.

3. See especially *The Book of Knowledge of Ingenious Mechanical Devices (Kitab fima rifat al-hiyal al-handassiya) by Ibn al-Razzaz al-Jazari*, translated and annotated by Donald R. Hill (Dordrecht, 1974), 170 ff: perpetual flutes. These employ trompes and trompe-like devices.

4. G. Hollister-Short, 'On the suction/lift pump', *History of Technology*, 1993, 15: 57–75, esp. 58.

5. E. McCurdy (ed.), *The Notebooks of Leonardo da Vinci* (London, 1938), 2: 421. A passage in the *Codex Atlanticus*, f.271 v.a., runs: 'With the help of the mill I will make unending sounds from all sorts of instruments which will sound for so long as the mill shall continue to move.'

6. Enzo Baraldi, 'Ordigni e parole dei maestri da forno Bresciani e Bergamaschi: Lessico della siderurgia indiretta in Italia fra XII e XVII secolo', in Philippe Braunstein (ed.), *La Sidérurgie Alpine en Italie (XIIe–XVIIe Siècle)*, Collection de l'École Francaise de Rome 290 (Rome, 2001), 163–213, but see especially sub Trombe: 'Del suo impiego come macchina soffiante aveva riferito della Porta del 1589 (*Magiae naturalis ..., Libro XIX, De pneumaticis*). In area appenninica si diffuse a partire dal XVII secolo ...', 205. Tivoli had a brassworks so blown only in 1653. Since the trompe is more commonly known as Catalan (as in Catalan Forge) it seems likely that transmission took place eastwards from that region. It should be noted also that the Catalan-type forges that Georges Richard studied in the 1830s (see note 7 below) were producing iron by the direct method, whereas Baraldi's study has to do with those producing iron by the indirect method.

7. J. Percy, *Metallurgy: The Art of Extracting Metals from their Ores ... Iron and Steel. Section II* (London, 1864), 289, quoting G.T. Richard, *Étude sur l'art d'extraire immédiatement le fer de ses minerais sans converter le métal en fonte* (Paris, 1839), 181–2, and *Atlas* (Paris, 1838), plate 2. For an extended discussion of the theory and practice of the trompe see Richard, 160–229. The trompes he investigated in the Ariège département were between 10 and 15 per cent efficient. This was of no account, however, given an abundant supply of water.

8. G.B. Aleotti, *op. cit.* (1), 88: 'Fabricare una stanza nella quale il tempo, che ci piacerà

sempre vi spiri vento, che la refreschi, e poco, e molto à voglia nostra' ['How to build a room in which the temperature shall always be comfortable by means of moving air which chills it a little or a lot according to your desire.']. A room cooled in this way was to be found at the Villa d'Este.

9. A. Keller, 'Pneumatics, automata and the vacuum in the work of Gian Battista Aleotti', *British Journal for the History of Science*, 1967, 3: 338–47. Keller is chiefly interested in examining the ambivalences in Aleotti's language as Aleotti seeks to explain the phenomena he is describing.

10. Nicholas Audebert's journal, 'Voyage d'Italie', British Library MS, Lansdowne 720, folios 311r–341v, *c.*1582, contain his description of the garden at the Villa d'Este. The passages describing the mechanical arrangements may be conveniently consulted in R.W. Lightbown, 'Nicholas Audebert and the Villa d'Este', *Journal of the Warburg and Courtauld Institute*, 1964, 27: 164–90, but see especially 181–5. At the Fountain of the Dragons (f.331r–f.331v) one could hear sounds of gun fire and of cannon: 'quattre dragons jettent l'eau par la bouche ... faict un pareil bruit que corps de harquebuses ... on entend parmy ce bruit comme quelques coups de gros canon.' At the Fountain of the Birds (f.332r–f.333r) on a metal tree were 20 small birds, each of which sang its own song so perfectly that one could not tell the difference between real birds and these of bronze. Then a mechanical owl appeared and the birds fell silent. When the owl withdrew, the birds began to sing again but not all at once. At first only a chaffinch sang ... 'non toutes foys tout a coup et tous ensemble, mais un seul commence le tintin d'un Pinson ...' Then, one by one, all the others joined in until the mechanism caused the owl to return yet again. Perhaps the masterpiece was the Fountain of the Organ (f.325v–f.337v) where, without any human intervention, an organ supplied with air through 22 pipes played a variety of tunes during which recitals a nightingale sang continuously. Of this Audebert writes: 'La fontaine des Orgues ... surpasses du tout les aultres ... pour l'artifice et secrets ingenieux qui y sont.' And all done by a Frenchman!

11. How these effects were produced, or some of them at least, can best be seen by referring to Gaspar Schott's engraving of a hydraulic organ which closely complements Audebert's description in most respects. See G. Schott, *Mechanica Hydraulico-Pneumatica* (Würzburg, 1657), 424, Iconismus, XLI. It is clear from the engraving that carillon technology, in the form of the programme drum, earlier developed in the Low Countries, had been married to the Heronic inventory. The earliest full description of how programme drums were produced is to be found in M.D.J. Engramelle, *La tonotechnie ou L'art de noter les cylindres, et tout ce qui est susceptible de notage dans les instruments de concerts mechaniques* (Paris, 1775), Engramelle's object was to expose the art (hitherto kept secret because it was a craft mystery) of how to 'point' (noter) cylinders for reproducing music mechanically.

12. G.B. della Porta, *Pneumaticorum Libri Tres* (Naples, 1601); S. de Caus, *Le Raison des Forces Mouvantes* (Frankfurt am Main, 1615); G. Branca, *Le Machine* (Rome, 1629).

13. For the text of Torricelli's letter to Ricci see G. Loria and G. Vassura (eds.), *Opere di E. Torricelli* (Faenza, 1919), 3: 186–8.

14. If von Guericke did ever possess copies of Aleotti's or della Porta's works, he would have read about the experiments with ramrods and cannons. However, we are not now likely to find out. The Magdeburg archives, among which were preserved the bulk of von Guericke's papers, were destroyed by Allied bombing in 1944.

15. G. Schott, writing in 1656, says of this device that von Guericke 'excogitavit is paucos ante annos machinam ...'.

16. G. Schott, *Experimentum novum Magdeburgicum quo vacuum aliqui stabilire, alii evertere conantur; inventum primo Magdeburgii a ... Othone Gericke* (Würzburg, 1657), 460–1. Here Schott, reporting the contents of von Guericke's letter of 22 July 1656, writes: 'Ait IV antliam pneumaticam etiam post longam agitationem adhuc posse agitari, sed tanto difficilius post aerem extractum, quanto gravior est aer circumstans, qui pistillum extractum suo pondere retrudit intus. Totam igitur in resiliendo intra antliam, adscribit ipse externi aeris gravitati ...' so that 'si antlia adhiberetur unam ulnam in diametro seu in latitudine aut amplitudine habens ... supra 1200 librarum pondus requisitum si ad pistillum extrahendum.'

17. N.R. Hanson, *Patterns of Discovery* (London, 1962), 5. European languages are full of fossil terminology quite apart from 'sunrise', 'lever du soleil', etc., for we still speak of 'suction' etc.

18. C. Huygens, *Oeuvres Complètes* (The Hague, 1950), 22: 241 ff, describes on 10

February 1673 how several months earlier the idea had suddenly come to him. In a letter to his brother, Lodewijk, of 22 September 1673 (*Oeuvres Complètes* (The Hague, 1897)), 7: 356–8, he described and sketched his gunpowder engine.

19. D. Papin, 'Nova Methodus ad vires Motrices validissimas Levi Pretio Comparendas', *Acta Eruditorum*, 1690, 410.

20. B.L. Add.MS.32504. For a discussion of this sketch see H.W. Dickinson, *Sir Samuel Morland. Diplomat and Inventor 1625–1695* (Cambridge, 1970), 74–80, and plates VIII and VIX.

21. T. Birch, *The History of the Royal Society of London for Improving Natural Knowledge from its First Rise* (London, 1756), 1: 295. On 19 August 1663 'Mr. Hooke showed the figure of an engine for determining the force of gunpowder by weight . . .', and later (302) on 9 September 1663 'Mr. Hooke brought in a scheme of the instrument for determining the force of gunpowder by weight, together with an explication thereof.' The drawing of the tester is to be found in the original *Register*, 2: 298.

Why the Chinese Failed to Develop A Steam Engine

KENT G. DENG

INTRODUCTION: THE ISSUE

It is commonly accepted, owing to the work of Joseph Needham and his team, that China (or the Chinese) once led the world in science and technology. Given the evidence so far, one can hardly reject this general outlook that was once a rather radical view back in the 1950s and 1960s when Needham began to make his career with *Science and Civilisation in China*. Granted that the Chinese made wonders in medieval times, it remains a big puzzle why and how they did not go any further in entering into an indigenous modernity. This is known as the Needham question or Needham's puzzle. So far, many hypotheses have been put forward to tackle this puzzle, ranging from environment and resource endowment to institutions and cultural values.[1] At the macro-level, there is almost no stone unturned by scholarly investigators.

Since the second half of the 1990s, though, there has been a shift from a Eurocentric mentality to something close to a Sino-centrism or Orient-centricism.[2] With the rise of this new school of thought, known as the 'California School', the Needham puzzle has become less attractive and less relevant to China's past because China no longer represents the case of development oddity in world history. Rather, Western Europe is now regarded as abnormal if not extraordinary.

Now, regardless of political correctness or incorrectness, as these two approaches are not compatible in so many ways, a tacit battle line has been drawn among historians: one is either with Western Europe or with China according to one's faith, curiosity, passion, empathy and so forth.

However, at the micro-level things look very different. One may look at small fractions of development in Western Europe or East Asia, avoiding the trap of taking a side yet contributing to the understanding of either Europe or Asia, or both.

METHODOLOGY AND SOURCE INFORMATION

For the current theme, notoriously 'normative' and 'counterfactual' as the title indicates, the methodology is derived from a Darwinian observation of the natural world that not all the early species were able to evolve to primates; and not all primates to *Homo sapiens*. Meanwhile, authentic Darwinism never strips off the rights for non-*Homo sapiens* to exist but 'Let It Be' as the Beatles' famous lyric implies. With this attitude, it is justifiable to examine why and how other species remained less changed from the common ancestor in relation to primates or *Homo sapiens*. In other words, true Darwinism imposes no value judgement. This no-value-judgement perspective also frees us from the afore-mentioned 'centricism' of either side. So, for the present study, as long as we reject the idea that there is a God-given 'natural course' for all societies, it is conceptually helpful and methodologically justifiable to compare different growth trajectories to see the differences by using one type (or its evolution) as a benchmark.

By the same token, in the technical world in the hands of the *Homo sapiens* there have been several long chains of evolution. There is little doubt that the development of the steam engine was at the end of one such chain which was specifically associated with England and then Western Europe.[3] In other societies, including much of Eastern Europe and all of Africa and Asia, this particular line of development was evidently absent.[4]

What this essay intends is to go through China's technological inventory to see why those necessary conditions for the steam engine to develop were not present. In doing so, we can turn a normative approach into a positive one.

In terms of information, the primary sources are from Chinese literature. An important part of the analysis has to be devoted to the Chinese way of thinking. Hence, some original works of Chinese philosophy are consulted. Given the nature of the topic itself, much of the technical information is drawn from *Exploitation of the Works of Nature* (*Tiangong Kaiwu*), a commonly recognized classic with rare illustrations, written in *c.* 1637; and, to a lesser extent, *Wangzhen Nongshu* (*Wang Zhen's Treatise on Agriculture*), an earlier book (1304) which included descriptions of farming equipment.[5]

CHINESE UNDERSTANDING OF ENERGY AND ITS FORMS

The secret of the working of the steam engine is to convert one form of energy into another form. The Europeans may have obtained the knowledge of energy inter-conversions by trial and error in the same process of making the first functional steam engine, and thus the story of Watt's kettle. Nevertheless, such knowledge was imperative for engine-making, although in reality such knowledge may well have been a product of an invention process in its own right.

Looking at Chinese knowledge, energy and its forms were understood in the form of 'positive and negative forces' (*yinyang*, literally meaning 'negative and positive') and 'five elements' (*wuxing*, literally 'five running fluxes'[6]). The whole knowledge package is known as *yinyang wuxing*, or 'five moving elements under the influence of negative and positive forces', for short 'five elements under two forces'.

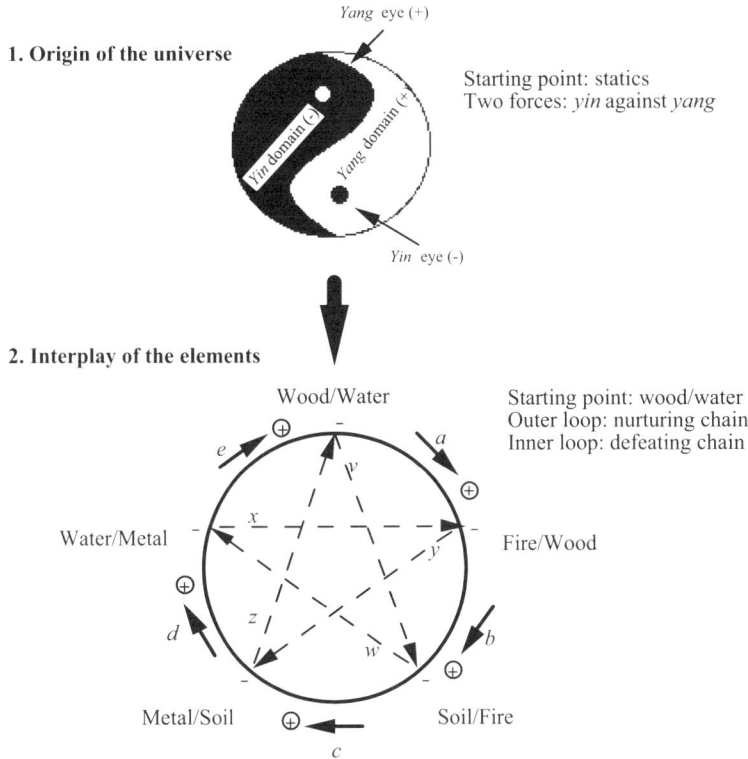

1. Origin of the universe

Yang eye (+)

Starting point: statics
Two forces: *yin* against *yang*

Yin eye (-)

2. Interplay of the elements

Wood/Water

Starting point: wood/water
Outer loop: nurturing chain
Inner loop: defeating chain

Water/Metal

Fire/Wood

Metal/Soil

Soil/Fire

Figure 1   Chinese understanding of energy and its forms, a standard view.

Source: (1) Dong Zhongshu, *c.* 104 BC, *Chunqiu Fanlu* (*Many Dewdrops of Spring and Autumn*), reprint, 1975, Beijing: Zhonghua Books, pp. 389, 392; (2) Wang Anshi, *c.* 1086, *Wangwengong Wenji* (*Collection of Wang Anshi's Works*), reprint, 1974, Shanghai: Shanghai People's Press, p. 281; (3) Zhou Dunyi, *c.* 1073, *Zhouzi Quanshu* (*Complete Collection of Master Zhou's Works*), publishers unknown, vol. 1, ch. 'Taiji Tushui' ('Illustrated Taiji'); also, (4) Wu Feng (ed.), 1990, *Zhonghua Sixiang Baoku* (*Essence of Chinese Intellect*), Changchun: Jilin People's Press, ch. 4.

Note: Positive and negative signs are indicators of (1) positive and negative matter-energies of *yin* and *yang*, and (2) the nurturing effect and defeating effect of the recipient element, respectively. During the Western Han Period (206 BC–AD 24), the starting point for the interplay among the five elements was wood. It became water under the Northern Song (960–1127).

Figure 1 represents a standard framework. The starting point is a static world (called *taiji*, literally 'the great origin') in a circle, before the two forces, *yang* (representing the positive but shapeless 'matter energy') and *yin* (representing the negative but shapeless 'matter energy'), kick in to work their way out of the static trap. The real dynamics lies in the small dots (called *yan* or 'eyes') in the opposite force domains that determine the curved line in the *taiji* pattern. The curved line shows the tension between the two domains. The product of this *yin–yang* intercourse or synergy is the first element of water (*shui*), which nurtures wood (*mu*, the second element), which in turn nurtures fire (*huo*, the third element), and so forth in a chain.[7] The nurturing loop is complete in a clockwise one-way traffic.

The mechanisms do not stop here. According to the *yin–yang* theory, the relationships between the five elements are not always friendly. The star-shaped routes, also clockwise, represent hostility between two elements: water defeats fire, fire defeats metal (called *jin*); metal, wood; wood, soil (called *tu*); soil, water. The Chinese material world is believed to be derived from the interplay of these five elements.[8]

In nature, the whole approach can be categorized as one of chemistry, not physics. Also, in nature, the model reaches a general equilibrium in terms of energy input and output. For example, if water goes to wood along arrow *a*, the energy input increases. If the increased energy flows in the inner ring along arrow *y* to soil, the increased energy will be reduced. But, the energy will recover if flowing in the outer ring along arrow *d* to metal. If energy flows freely and endlessly between the two rings/loops and among five elements, one gets a total balance sheet of an energy account. Based on this assumption, to make more energy out of the system will be impossible.

The origin of the Chinese *yin–yang* theory began around the Spring and Autumn Period (770–476 BC) and became popularized amongst Taoists and Confucians who adopted the idea for philosophical and political purposes. By the Song Period, after some fifteen-centuries of evolution, it was perfected in the hands of the Confucian literati. In the process, many thinkers left their mark.[9] This 'five elements under two forces' knowledge package remained unchallenged in China until the visit by the Jesuits who introduced post-Renaissance sciences to the Ming Empire.

The implication of the whole Chinese energy system is predictably of a general equilibrium for energy input and output: any growth generated in the outer ring is to be cancelled by hostility in the inner ring, and vice versa. The whole layout is cleverly maintained with its own logic. The Chinese system is a closed circuit with a strong dependency on natural endowments. In other words, human impact is considered trivial if not completely nonexistent or irrelevant. This differs fundamentally from the Western approach of applied physics and engineering which emphasizes how efficient energy input can produce an output, hence the invention of the pressure boiler and barometer.

However, it would be unfair to suggest that after the Song the package was completely frozen. A new version was created by Song Yingxing (1587–?) as shown in Figure 2. The novelty of Song's new interpretation is three-fold: (1) the movement is anti-clockwise; (2) both the outer ring and inner ring represent defeating chains; (3) it leads to a general disequilibrium or a total collapse, as any input of energy will diminish. For example, if water goes to metal along arrow *a*, the energy input reduces. If the reduced energy flows either in the outer ring along arrow *b* to soil, or in the inner ring along Arrow *x* to wood, it reduces once again.

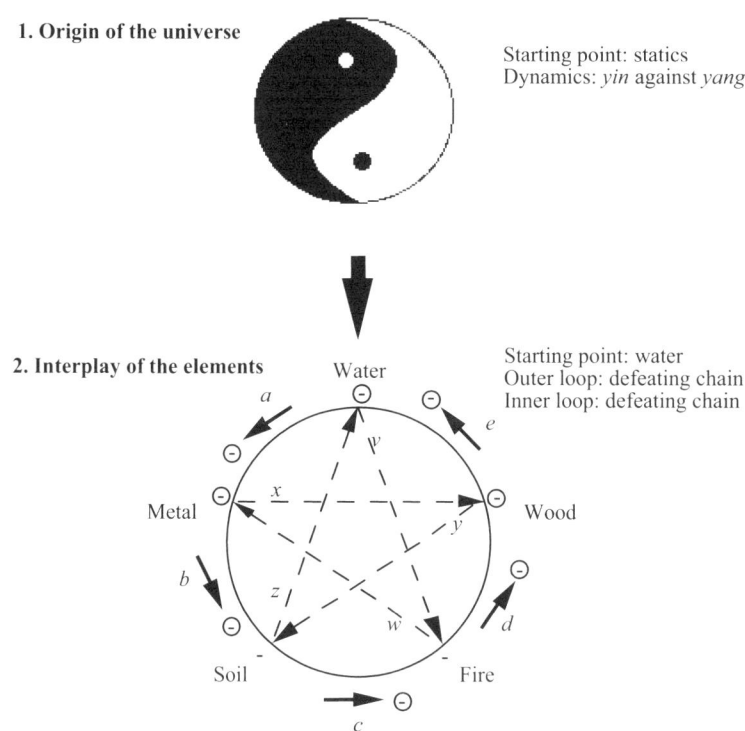

1. Origin of the universe

Starting point: statics
Dynamics: *yin* against *yang*

2. Interplay of the elements

Starting point: water
Outer loop: defeating chain
Inner loop: defeating chain

Figure 2   Chinese understanding of energy and its forms, a minority view.

Source: Song Yingxing, 163?, *Yeyi Lunqi Tantian Silianshi* (*Free Comments on Energy, Heaven and Compassion*), reprint, 1976, Shanghai: Shanghai People's Press, pp. 80–1.

Note: Negative signs are indicators of the defeating effect on the recipient element.

This is a pessimistic approach. What this tells us, however, is that Chinese thinking was trapped in a circularity.

Consequently, the applied 'five elements under two forces' theory was concentrated in four areas: (1) Chinese philosophy due to its evolutionary outlook, (2) Chinese calendar sequence (called *ganzhi*, meaning 'Heavenly Stems and Earthly Branches') thanks to its cyclical outlook, (3) Chinese medical and *fenghui* reasoning (literally 'art of wind and water fluxes') owing to its equilibrium outlook, and (4) Chinese fortune-telling because of the amount of probabilities offered (see Figure 3). So, if one uses post-Renaissance attitudes as a benchmark, the Chinese 'five elements under two forces' approach is on a very different developmental path. The point here is that in character the European and Chinese approaches are mutually exclusive.

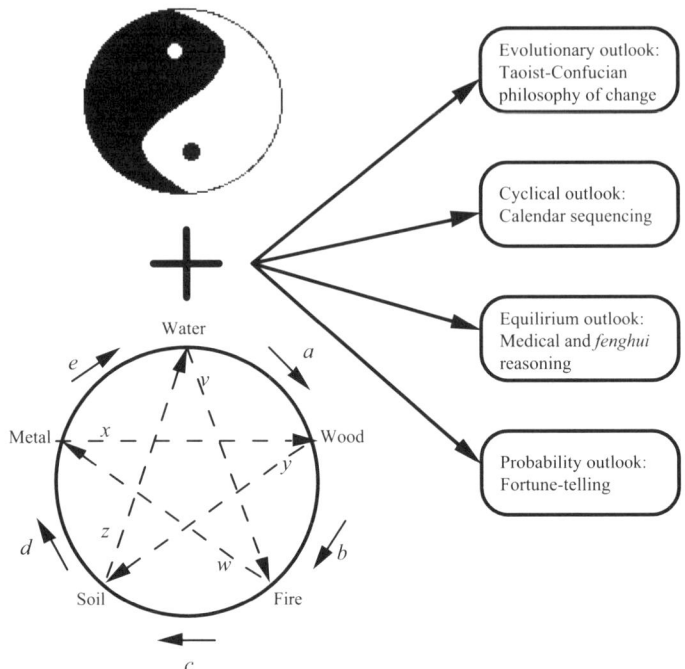

Figure 3   Applied 'Five Elements under Two Forces' theory.

Source: Yang Li., 1999, *Zhonghua Wuqiannian Wenhua Jingdian* (*China's Cultural Classics for 5,000 Years*), vol. 1, Beijing: Beijing Science and Technology Press, chs. 1–19, 42–7.

CHINESE USE OF WATER AND FIRE AS ENERGY SOURCES
WITHOUT STEAM

The steam engine represents a concept as well as a specific way to use non-human energy to produce goods and services. Although the concept of non-human energy was not new by the time the first functional engine was constructed, the specific solution was.

The steam engine itself has two components. One is what we call the 'hardware', i.e. the machine itself. The other is the 'software', composed of abstract scientific principles and technical knowledge. What is so unique about the engine is that it is a carefully designed machine whose function is based on an artificially controlled environment at all times: temperature, pressure, movement and so forth. This could not be achieved by sheer trial and error with purely visual observation, which was the norm for the emergence of any premodern technology. What on the Chinese cards was close to such an engine? We begin with steam as a form of energy.

The Chinese were undoubtedly keen on the use of water and fire, both belonging to the five elements, as energy sources. Water as an energy force

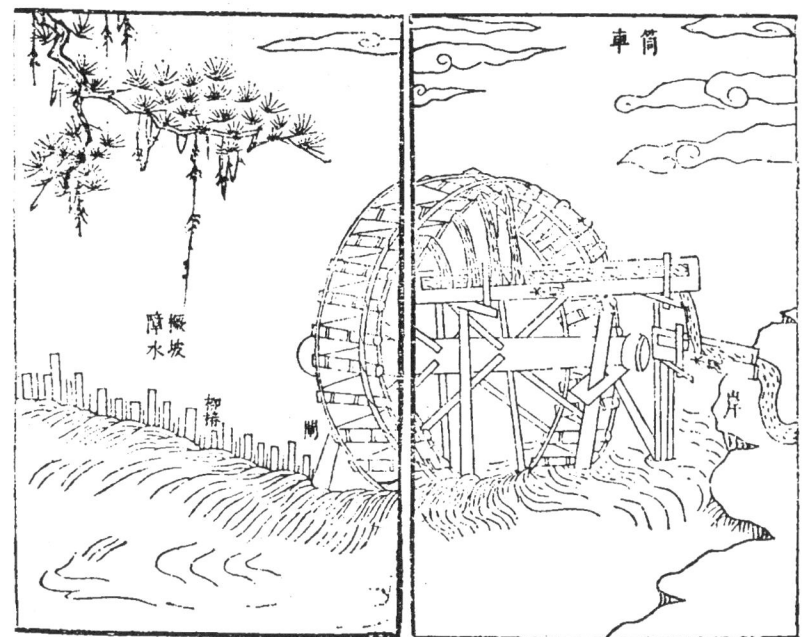

初刻本插图五：筒车。

Figure 4 Traditional scoop wheel for irrigation.

Source: Song Yingxing, 1637, *Tiangong Kaiwu* (*Exploitation of the Works of Nature*), reprint, 1978, Hong Kong: Zhonghua Books, ch. 1: Entry 'Shuili' ('Irrigation').

is relatively common in South China for irrigation and food processing. The best examples were the scoop wheel and the water-powered mills, as shown in Figures 4 and 5. Both were available from the Yuan Period (1271–1368) on. Although they were as sophisticated and efficient as a premodern machine could be, their spread was narrowly constrained by suitable water sources.

By the Ming Period (1368–1644), fire made from wood/charcoal, coal and natural gas was used in brine-processing for salt (see Figures 6 and 7), brick-making (Figure 8), ceramics-baking (Figure 9), and metal-casting (Figure 10). But, there was no sign that the Chinese went beyond that.

Considering that fire and water are the central elements for the concept of the steam engine, in China (or in other traditional societies) there was no case in which these two were combined to produce energy for production purposes. In particular, there was no attempt to convert one form of energy to another.

Figure 5   Multiple water-powered mills.

Source: Wang Zhen, 1304, *Wangzhen Nongshu* (*Wang Zhen's Treatise on Agriculture*), reprint, 1981, Beijing: Agriculture Press, pt. 3 'Nongqi Tupu' ('Illustrations of Farming Tools').

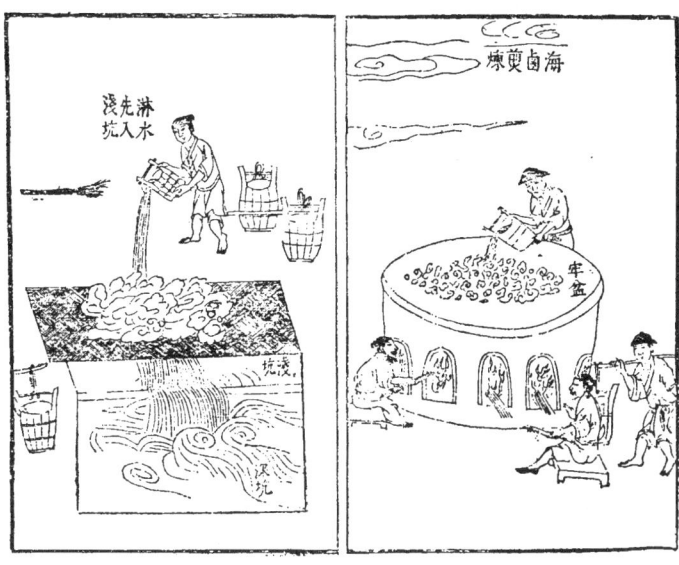

初刻本插图四〇：淋水先入浅坑。　　初刻本插图四一：海卤煎炼。

Figure 6 Brine-processing with firewood.
Source: Song, *Works of Nature*, ch. 5: Entry 'Haishui Yan' ('Brine for Salt').

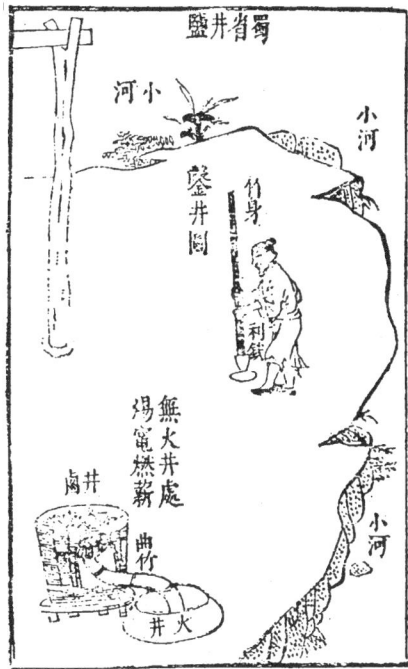

Figure 7 Brine-processing with natural gas.
Source: Song, *Works of Nature*, ch. 5: Entry 'Jing Yan' ('Well Salt').

煤炭烧砖窑

初刻本插图五〇：煤炭烧砖。

Figure 8 Brick-making with coal.
Source: Song, *Works of Nature*, ch. 7: Entry 'Zhuan' ('Bricks').

Figure 9 Ceramics-baking with firewood.
Source: Song, *Works of Nature*, ch. 7: Entry 'Ying Wong' ('Vases and Jars').

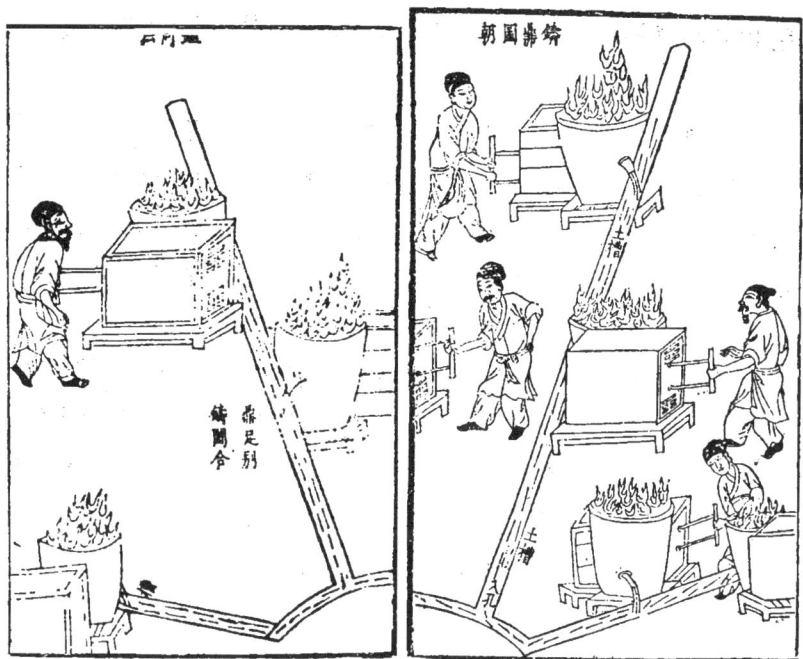

Figure 10 Furnaces for metal-casting.

Source: Song, *Works of Nature*, ch. 8: Entry 'Ding' ('Cooking Vessels').

CHINESE MACHINES WITHOUT AN ENGINE

The Chinese were undoubtedly capable of inventing a range of machines, some rather heavy, for production purposes. However, without exception, these machines depended on muscle power or crude natural forces such as water and wind. What they had in common was the absence of a man-made engine. Good examples of the Chinese approach can be seen in Figures 4 and 5 above. Figures 11 and 12 illustrate further the Chinese technological achievement in sea-going ships, powered solely by wind. The technology was sophisticated and adequate for events such as the historical voyages made under Admiral Zheng He of the Ming.[10]

Obviously, the exploitation of the crude natural force of wind is also confined by locations. To transcend such a constraint, the Chinese went for muscle, human or animal (see Figure 13). It is not surprising that the use of muscle was the predominant factor among all Chinese machines.

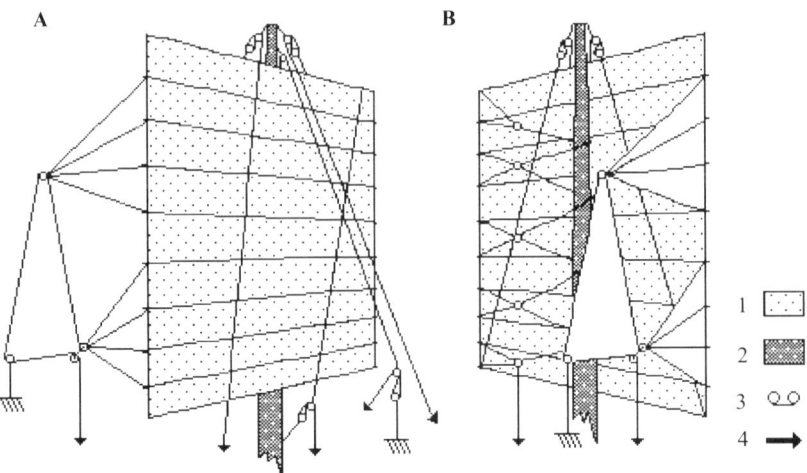

Figure 11   Sail design of a Chinese sea-going ship.

Source: Based on information from (1) Zhou Shide, 1963, 'Zhongguo Shachuan Kaolue' ('On the Shallow Water Ships in China'), *Kexueshi Jikan* (*Collected Works on History of Sciences*), no. 5, p. 44; (2) G.R.G. Worcester, 1971, *The Junks and Sampans of the Yangtze*, Annapolis [Md.]: Naval Institute Press, pp. 75–85, 163, 174–5; (3) Joseph Needham, 1971, 'Civil Engineering', *Science and Civilisation in China*, vol. 4, pt. 3. Cambridge: Cambridge University Press, p. 596 and figs 1010–19; (4) K.C. Danforth (ed.), 1982, *Journey into China*, Washington DC: National Geographic Society, pp. 162, 276–7, 280, 482–3; (5) Douglas Phillips-Birt, 1962, *Fore and Aft Sailing Craft*, London: Seeley, Service and Co., fig. 14.

Note: A–forward view; B–stern view; 1–sail (suspended on the yard, it is folded or unfolded with the vertical control of the battens); 2–mast; 3–pulley blocks; 4–rigging.

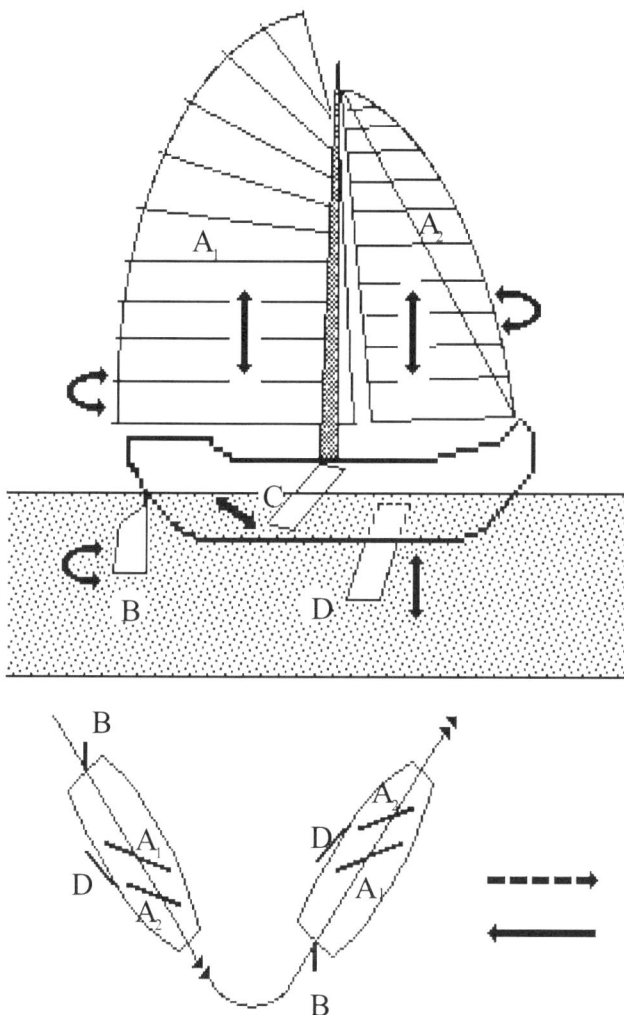

Figure 12   Control devices in Chinese ships.

Source: Based on Zhou, 'On the Shallow Water Ships in China', pp. 42–4; see also Needham, 'Civil Engineering', p. 593.

Note: The top figure is a horizontal view. The bottom figure is a vertical view of tacking into the wind with the steering and direction devices. Symbols: A_1 and A_2–mainsail and foresail; B–stern rudder; C–leeboard; D–centreboard. The two-headed arrows indicate control directions; double arrows indicate course for the ship; solid arrow indicates wind direction; arrow with a broken tail indicates intended direction.

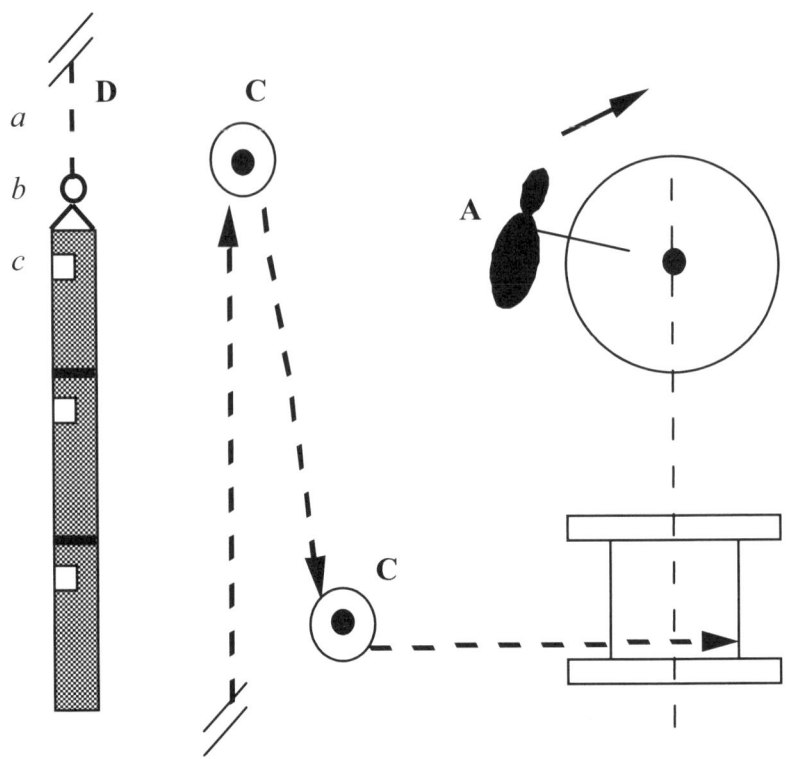

Figure 13   Animal-powered lifting device during the Ming.

Source: Based on Song, *Works of Nature*, ch. 5: Entry 'Jingyan' ('Brine Well').

Note: A–draught animal; B–capstan; C–pulleys; D–water container made of a bamboo pipe; a–rope; b–hook; c–bamboo pipe with openings. Solid arrow – the way the animal is heading; broken arrows–the direction in which the rope moves.

To tap more power, the Chinese utilized larger and larger water wheels, taller and taller sails and more and more draught animals (for animals see Figure 14). But they never escaped from the 'traditional power trap' of crude natural forces and muscles.

It has been widely accepted, however, that some of the Chinese inventions possessed some suggestive/speculative significance in terms of a possibility of developing a mechanical engine (be it a steam one). The only evidence so far is from the alleged design of the Chinese mechanical bellows (see Figure 15): it has a piston, it is linked to a wheel, and it moves. The only element absent was to reverse the energy input direction and let the piston drive the wheel. Voilà!

初刻本插图六八：合挂大车。

Figure 14 A heavy cargo vehicle with eight draught animals.

Source: Song, *Works of Nature*, ch. 9: entry 'Che' ('Vehicle').

There are several problems in the possible 'conversion' of the Chinese bellows to a steam engine. First, the whole design was to move air of a normal pressure, not steam of a high pressure. To convert it to use steam requires radical changes in the materials used to withstand a high pressure at the very least. Second, the cylinder has to be air tight, a skill that was absent in China. Third, the energy input-to-output ratio in its current form determines that to reverse the motion (so that the piston becomes an active, energy-donating component while the waterwheel becomes a passive, energy-recipient component) is unworkable. All these problems are deeply rooted in one thing: the absence of the concept of 'engine' (a prime mover).

In essence, the function of bellows (a machine tool) itself produces no vacuum in any part of the machine or its movement. At most, a bellows uses air pressure which is determined by the required function of a pumping machine (a machine tool). On the contrary, it is the vacuum that plays the central part in the steam engine (a prime mover). In other words, a steam engine and a bellows are mutually exclusive. If a machine can produce a vacuum, it cannot be at the same time a bellows, and vice versa. So, in principle, a steam engine and a bellows are not convertible to each

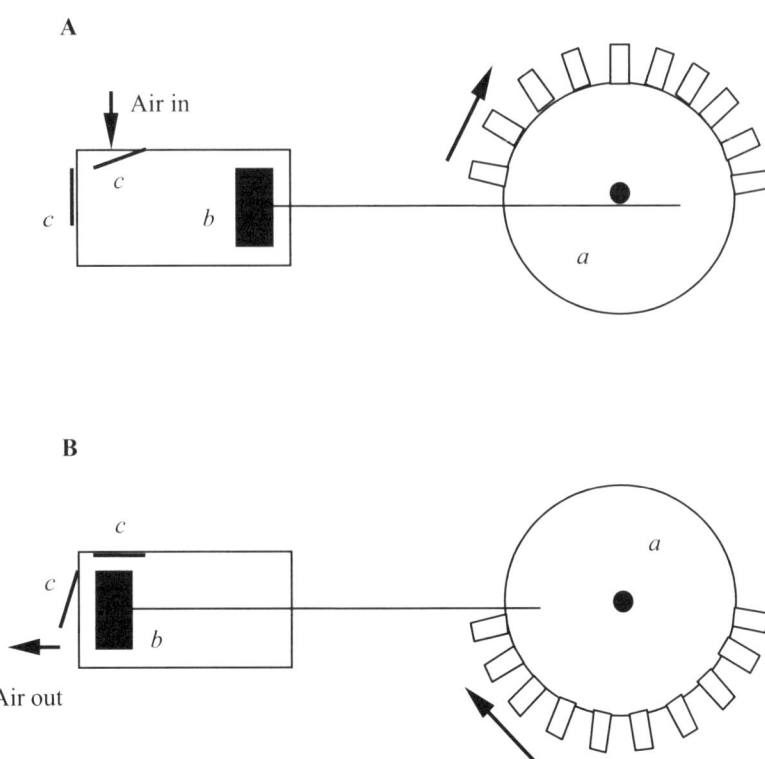

Figure 15 Chinese design of water-powered mechanical bellows, imagined.

Note: A–The first stroke; B–the second stroke; *a*–water wheel; *b*–piston; *c*–air valves. Arrows–directions to move in.

other. Such a fundamental conceptual difference has been long ignored as most scholars see similarities in the physical movement between a steam engine and a bellows.

In this context, it is worth noting that the ancient Chinese possessed the knowledge of atmosphere pressure to make siphons and pumps work as early as the fifth century.[11] They also understood and created a vacuum as a natural phenomenon and used it regularly in their medical practice known as cupping (*ba huoguan*, literally 'using fire [to create a vacuum] to suck the pot').[12] However, there is no evidence that the Chinese went any further to create a prime mover with such knowledge, which itself remained at a primitive stage throughout Chinese history.

This is not all. A closer look at what was illustrated in *Wang Zhen's Treatise on Agriculture* reveals a machine without a piston. Instead, the bellows were made of an open fan (see Figure 16). This raised serious doubt whether the Chinese ever used the piston at all. Without the piston, the Chinese were even further away from an 'engine'.

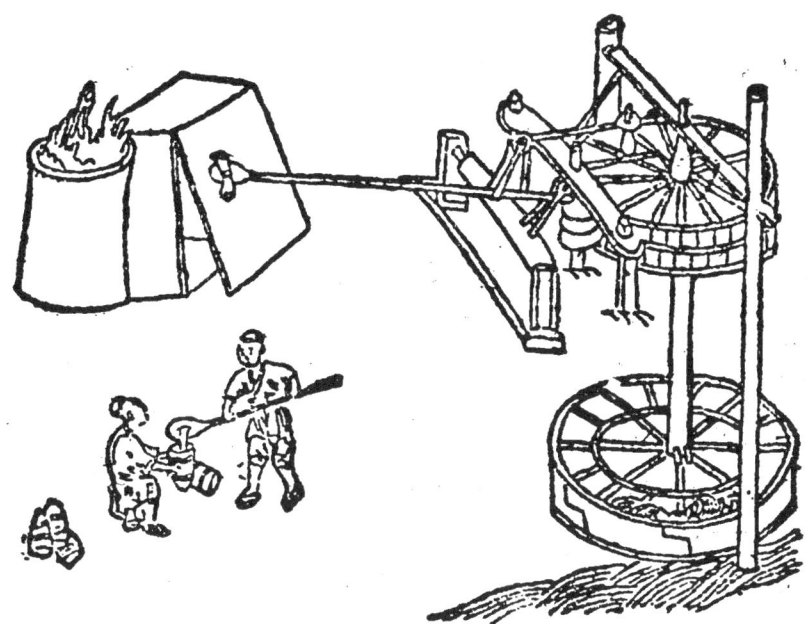

Figure 16   Chinese water-powered bellows, actual.

Source: Wang, *Treatise on Agriculture*, pt. 3 'Nongqi Tupu' ('Illustrations of Farming Tools').

WHY MINING WAS NOT THE HOTBED FOR THE STEAM ENGINE IN CHINA

It has been argued that the main reason that the Chinese did not invent the steam engine was the lack of demand for such a device in the Chinese mining industry. In particular, coalmines in North China were largely free from flooding.[13] This argument is undoubtedly inferred from the British experience in the late eighteenth century. It is now commonly enough claimed that the Roman conquerors in England faced the same problem of underground flooding but never came to the point of inventing the steam engine. This strongly suggests that the British invention of the steam engine in the eighteenth century was rational but not inevitable; accidental, but not predestined or teleological. To follow this line and suppose that China had the need to pump water from some of its coalmines (or any mines), what was on the cards realistically?

The Chinese pumping technology developed along two lines. One line is shown in Figure 13, a water-lifting technology which was characterized by its discontinuous motion. This would cope with slow underground flooding. The other line is a pump with continuous motion. There was an existing model in paddy irrigation (see Figure 17). There was no reason why these two types were not suited for underground water (see Figure 18). Arguably, these two types were adequate for small mines of the non-capitalist type. In

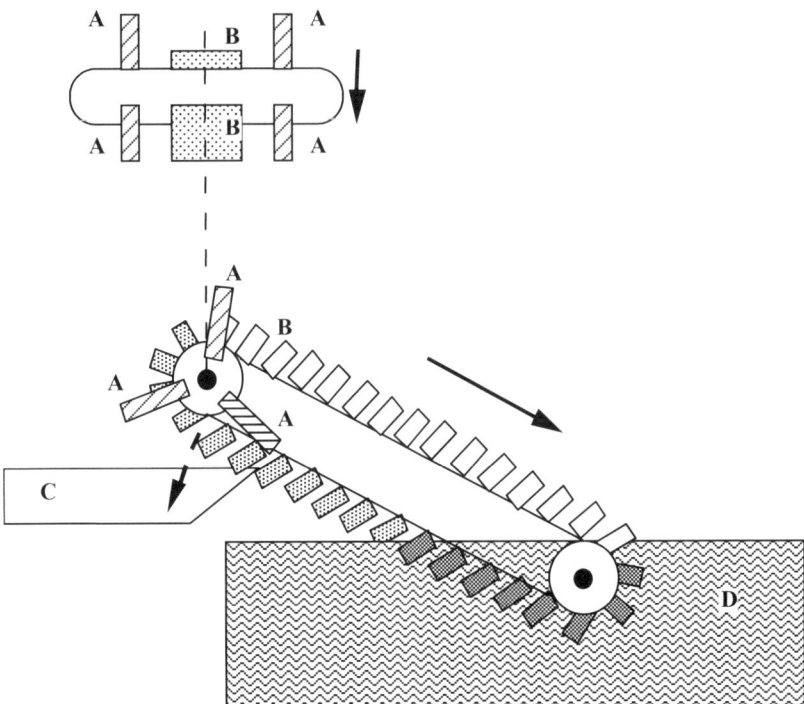

Figure 17   Man-powered 'dragon-bone water pump' for paddy irrigation

Source: Based on Song, *Works of Nature*, ch. 1: Entry 'Shuili' ('Irrigation').

Note: A–Pedals; B–water scoops; C–ditch; D–water source. Solid arrow–the direction to move in; broken arrow–where the water moves.

this context, given that there was plenty of coal and water in China, the need for pumping underground water *per se* was at best a necessary but not a sufficient condition for the steam engine to be invented.

What seems to have happened was that the Chinese followed a 'Ricardian expansion' mode in their coal mining and avoided mines with the underground flooding problem altogether. The evidence is from the most celebrated travels entitled *Master Xu Xiake's Travels (Xuxiake Youji)* of the Ming Period.[14] Xu's lifetime pursuit was visiting caves and mines (for over 30 years, 1607–40). But, he never came across a mine in wet conditions. This was compatible with the lack of dependency on coal in premodern China.

Even taking into account that coalmines in China are more vulnerable to gas and combustion than flooding, one still cannot claim that the lack of flooding alone was the reason for the non-development of the steam engine there. This is because for ventilation or fire prevention, the use of the steam engine – a machine as versatile as is – would be indeedhandy. But this

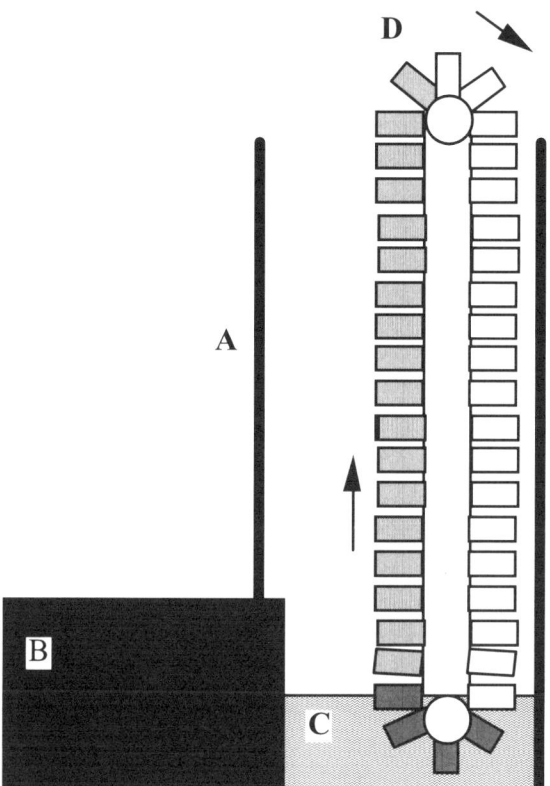

Figure 18 A concept water pump for Chinese mines.

Note: A–Shaft; B–Coal deposit; C–underground water; D–water pump. Solid arrow–where the water moves.

development was rather late even in Britain, the homeland of steam-engine technology. There was no sign that the Chinese ever moved in the direction. Instead, for example, they used bamboo pipes for ventilation (see Figure 19).

Overall, the drive for mass production was absent in China. Given the limited scale of industrial operations, the traditional, low-cost and pragmatic technology was best suited for China. This applies to not only coal mining but also the economy in its totality.

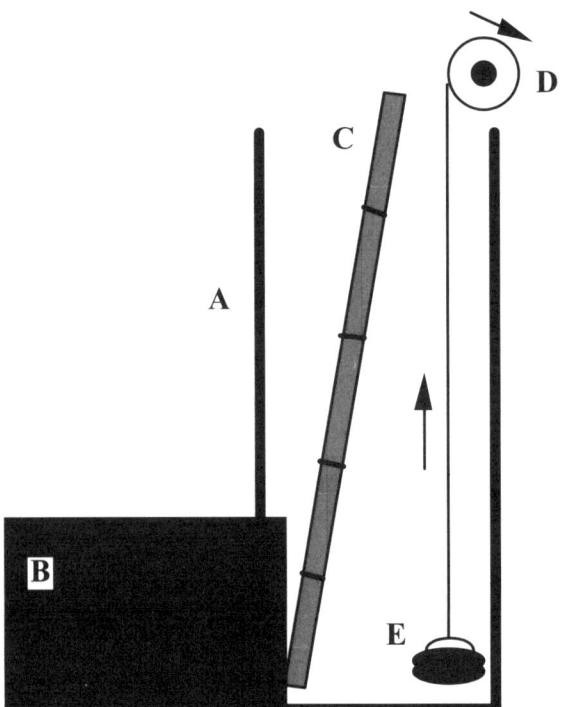

Figure 19   Gas ventilation with bamboo pipes.

Source: Based on Song, *Works of Nature*, ch. 11: Entry 'Meitan' ('Coal').

Note: A–Shaft; B–coal deposit; C–bamboo ventilation pipe; D–lifting device; E–coal container. Arrows–the direction of motion.

FINAL REMARKS

The reasons why premodern China did not invent the steam engine were multiple. At the level of abstract reasoning and scientific thinking, the Chinese were firmly confined within the boundary of the *yin–yang* forces and the five basic elements. The thought became perfected and sophisticated but never modern. At the more mundane and practical level, the Chinese did not go beyond the exploitation of crude energy force (such as water, wind and animal/human muscles), despite the widespread knowledge of labour-saving devices.

All the results from present analysis strongly suggest that what were completely absent in premodern China were the concept of energy conversion and the concept of engine, both central to the British Industrial Revolution. The steam engine was both a cause and an effect of mass production. Without mass production such an engine is not viable economically even in the short run. Therefore, the ultimate reason has to be found outside the domain of technology itself.[15] In this context, even if the Chinese had been given 'sufficient time', the invention of the steam engine would almost certainly have had no chance to take place.

Notes and References

1. See Kent G. Deng, 'A Critical Survey of Recent Research in Chinese Economic History', *Economic History Review* LIII, 1, February 2000, 1–28.

2. See e.g. B.R. Wong, *China Transformed, Historical Change and the Limits of European Experience* (Ithaca and London, 1997); A.G. Frank, *ReOrient: Global Economy in the Asian Age* (Berkeley, 1998); Kenneth Pomeranz, *The Great Divergence, Europe, China and The Making of the Modern World Economy* (Princeton, 2000).

3. As for such an invention becoming a watershed that changed the trajectory of the society, which nurtured the chain in the first place, it is another matter, which is beyond the scope of this paper.

4. I deliberately avoid the term 'missing' and use 'absent' as the former suggests that there was a probability for the particular development to take place. To suggest such a probability is fatalistic and non-Darwinian.

5. The author of the first book was Song Yingxing who lived 1587–164?. The dates of the second author are unknown.

6. The origin of the term *xing* (meaning 'moving' and 'travelling') was 'samskara' in Sanskrit. The concept was borrowed from the Buddhist concept of an endless change.

7. The Chinese terminology is *xiangsheng*, meaning 'to energize and vitalize'.

8. The Chinese terminology is *xiangke*, meaning 'to suppress and stifle'.

9. The most famous names include Dong Zhonshu (179–104 BC), Li Quan (*c.* seventh century), Liu Zongyuan (AD 773–819), Tan Qiao (*c.* tenth century), Shao Yong (1011–77), Zhou Dunyi (1017–73), Zhang Zai (1020–77), Wang Anshi (1021–86), Cheng Hao (1032–85), Cheng Yi (1033–1107), Zhu Xi (1130–1200), Lü Zhuqian (1137–81), Cai Shen (1167–1230), Zhen Dexiu (1178–1235), Su Tianjue (1294–1352), Li Ji (1311–75), Cao Duan (1376–1434), Wang Tingxiang (1474–1544), Lü Kun (1556–1618), Huang Zongxi (1610–95), Zhang Lüxiang (1610–74), Fang Yizhi (1611–71), Zhang Ying (1637–1708), Yang Mingshi (1661–1737), Ji Yun (1724–1805), Yu Yue (1821–1907), Yan Fu (1854–1921), Kang Youwei (1858–1927).

10. See Joseph Needham, 'Civil Engineering', *Science and Civilisation in China*, 4, pt. 3 (Cambridge, 1971); Louise Levathes, *When China Ruled the Seas: The Treasure Fleet of the Dragon Throne, 1405–1433* (New York, 1994); Gavin Menzies, *1421, the Year China Discovered the World* (London, 2002).

11. See Dai Nianzu, 'Mechanics', in Chinese Academy of Scieces (ed.), *Ancient China's Technology and Science* (Beijing, 1986), 132–4.

12. Ge Hong (*c.* 281–341) is commonly regarded as the first writer who mentioned the practice of cupping in China's medical history; see Chen Shikui and Cai Jingfeng, *Zhongguo Chuantong Yiyao Gailan* (Survey of Traditional Medicines in China) (Beijing, 1997).

13. Pomeranz, *The Great Divergence* (2), 65.

14. Xu Xiake, *Xuxiake Youji* (Master Xu Xiake's Travels) (Shanghai, 1982) reprint.

15. See Gang Deng, *The Chinese Premodern Economy – Structural Equilibrium and Capitalist Sterility* (London and New York, 1999), chs. 2, 3 and 6.

Response to Kent Deng[1]

DAVID WRIGHT

My response will deal with the final part of Kent Deng's paper, concerning the factors which led to the development of steam engines in China during the nineteenth century.

China had had a long history of technological innovation, using devices which extracted energy from wind, coal, wood and natural gas, as well as human or animal muscle-power, and an ancient tradition of metallurgical technology.[2] By 1800 (and perhaps long before) the prerequisites for the development of steam power were already present in China; by the mid-century the Chinese were building their own steam-powered vessels.

The earliest Chinese contacts with Western steamships were probably around Guangzhou (Canton) in the 1830s, and during the First Opium War (1839–42). Various wealthy merchants and officials saw the potential of these vessels, and the earliest steam prototype in China seems to have been built by Ding Gongzhen 丁 拱 振 (c. 1800–c. 1875) about 1843, although little is known about it.[3] We are fortunate, however, to have an eye-witness account of a slightly later vessel built by Pan Shicheng 潘仕 成 in the grounds of his garden in Guangzhou.

The eye-witness was Dr Aurelius Harland (1822–57), who had arrived in Hong Kong in 1847. The Harland family had a deep interest in steam power: Aurelius's father, Dr William Harland (1786–1866), a friend of George Stephenson (1781–1848), was himself a pioneer of steam-powered cars, whilst Aurelius's younger brother Edward (1831–95) went on to found the shipbuilding firm of Harland and Wolff.[4] Aurelius learned Chinese in Hong Kong, and seems to have been a significant contributor to *Bowu xinbian* 博物 新 編 [A new compilation on natural philosophy] (Guangzhou, 1854–5), edited by another British physician Benjamin Hobson (1816–73). *Bowu xinbian* would later play its own role in the transmission of steam technology.[5]

In one of his letters to his father, Aurelius described a steamship built by an un-named Chinese engineer who had worked on Western steamships as a stoker, second mate and finally an engineer. On his return he had persuaded Pan Shicheng to undertake the building of a Chinese steamship:

> The work is going on in Pontingqua [Pan Shicheng]'s private garden about 3 miles above the city , and there I can assure you I spent one of the most delightful afternoons since I left home, indeed it strongly reminded me of home and being so congenial to my own natural taste I could have staid [sic] till the boat was finished. The gardens

are very spacious but walled all round except at the gate where a small canal enters from the river to supply the fishponds: it is by the side of this canal that the shed containing the embryo steam boat is erected. The boat itself is 70 feet in length and 15 feet beam, built exactly in the Chinese form as no other would yet be allowed – all the machinery including boilers etc. is on deck, the paddles are principally of wood 10 feet in diameter and about $2\frac{1}{2}$ in width, very beautifully made, and strongly fastened with iron; there is only one cylinder of a foot diameter inside and a 3 feet stroke. It is laid on deck horizontally; but being nearly two feet below the level of the paddle shaft will work with great loss of power, for when the cam is at the highest point there will be a large portion of portion [perhaps 'portion of power' is meant – DW] wasted in the tendency to tear up the beams which contain the groove in which the cross bar at the head of the piston slides. The boilers are 4 in number, placed side by side in one furnace, each of them is 12 feet long and $2\frac{1}{2}$ feet in diameter, with a fire flue of 8 ins through each. They are made of brass hooped with iron, and the brass plates are brazed instead of being riveted together. As there is only one cylinder and no fly wheel, the motion will be very intermitting [sic], and the starting will be very difficult for there is nothing but the momentum of the vessel to carry the crank past the line of centre when the piston reaches each extremity of the stroke. The man [i.e. the Chinese engineer] is very intelligent and speaks good English – I pointed out some faults and improvement[s] which he said he would adopt saying he hardly expected to succeed at first but hoped to make a good one after 3 or 4 attempts when the workmen had a little more experience – he expects to have all this completed about a fortnight from this time.[6]

These pioneering efforts do not seem to have led to any large-scale shipbuilding programme, for it was not until the early 1860s when the Chinese court, smarting under internal attacks by Taiping rebels and their defeat by Western powers in the Second Opium War (1856–60), encouraged the policy of 'Self-Strengthening'. The scholar-general Zeng Guofan 曾國藩 (1811–72), who led the campaign against the Taiping rebels, asked two scholars Xu Shou 徐壽 (1818–84) and Hua Hengfang 華衡芳 (1833–1902), who were renowned for their knowledge of science and mathematics, to build a steamship at the Anqing 安慶 Arsenal. Xu and Hua, using information from *Bowu xinbian* and the Western steamships to which they had access, built a vessel on which Zeng and his family later travelled down the Yangzi River.[7] By the mid-1860s the Jiangnan Arsenal was established in Shanghai, where a series of steamships were built with the help of foreign artisans.[8]

The factors which, I suggest, finally led to the introduction of steam power into China by the 1860s were (i) access to examples of Western steam technology, (ii) official recognition of the military importance of steamships, (iii) official willingness to invest in the technology, (iv) the existence of individuals with the necessary knowledge, skill and perseverance to fulfil the project, (v) awareness of the scientific principles underlying steam power, (vi) sources of coal and iron ore, and the means of smelting the iron ore into steel, (vii) the organization necessary to ensure the coordination of the various processes, (viii) the will to overcome the objections of conservative or reactionary forces to the new technologies and, perhaps most importantly, (ix) military and economic pressures from foreign powers which gave urgency to the need to 'catch up'. Despite this sense of urgency amongst a small group of modernizing officials, it took several decades and the most crushing humiliation of all – defeat by Japan in 1895 – finally to convince the highest levels of the ruling elite that China had to adopt the new technologies wholeheartedly if it were to survive in the modern world.

Notes and References

1. For the general issues involved, see Ssu-yü Teng and John K. Fairbank, *China's Response to the West* (Cambridge, Massachusetts, 1979) and my *Translating Science* (Leiden, 2000).

2. See Joseph Needham *et al.*, *Science and Civilisation in China*, 7 vols.(Cambridge, 1956–), for the most comprehensive account of Chinese science and technology.

3. See Gideon Chen, *Lin Tse-hsü: pioneer promoter of the adoption of Western means of maritime defense in China* (Beiping, 1934), 36–7 and 46.

4. See the website www.geocities.com/razgbr/SirEdwardHarland.html

5. See the letter from Aurelius to his father William Harland, 24 April 1849, held in the Wellcome Institute for the History and Understanding of Medicine (classmark MS 7682/46) and *Translating Science* (1), 263–6.

6. Letter from Aurelius Harland to his father dated 24 June 1847, classmark MS 7682/32/1/2. Spelling and punctuation are as in the original. This letter is quoted by permission of Dr Erasmus Harland and the Wellcome Institute.

7. See John Fryer, 'An account of the Department for the Translation of Foreign Books at the Kiangnan Arsenal', in *North-China Herald*, 29 January 1880, 77–81.

8. See Meng Yue, 'Hybrid Science *versus* Modernity: the practice of the Jiangnan Arsenal, 1864–1897', in *East Asian Science, Technology and Medicine*, 16 (1999), 13–52.

Response to Kent Deng

CHUN-YU LIU

As Deng rightly points out, investigating why China had never invented a steam engine is itself counterfactual, and, to me, doomed to be beyond the explanatory power of any explanation from technology. On the methodological level, few would dispute that 'it is conceptually helpful and methodologically justifiable to compare different growth trajectories ... by using one species (or its evolution) as a benchmark'.[1] However, it should also be noted that a value-freed Darwinist approach does not necessarily free one from the so-called 'trap of centrism'. This is particularly so as moral and ethical issues and value judgements had long been indivisible parts of premodern Chinese scientific thought. One cannot understand the intimate interactions of cultural values and scientific activities without setting oneself within the specific intellectual and cultural context of pre-Jesuit China. Differing from the European tradition, concepts of morality and ethics in traditional China had integrated tightly with a value-freed or -neutralized natural world. Under the principle of the unity of nature and humanity 天人合一, Chinese scholars (especially after the Song period) associated the Confucian concept of benevolence (or ren 仁) with the Daoist metaphysical concept of Dao 道 and universe.[2] Such a theorization attached the nature of human reason to the phenomena and laws of the natural world, and injected at the same time moral and ethical qualities or elements into the natural order. For instance, the Song scholar Wen Tian-Xiang 文天祥 (1238–82) wrote 'There is the qi 氣 of righteousness among the Heaven and the Earth; such qi transmutes itself into abstract forms and flows among all matters of being.'[3] In other words, Chinese intellectuals up to the late Ming period were operating in a context of 'nature' that differed substantially from that of the Europeans. Only the literati saw the wholeness of nature, ethics and humanity not as a burden of knowledge but an inborn and requisite integrity.[4] Science and technology was incomplete by itself in pre-Jesuit China. It was never an end of its own, but, as Deng also recognizes, a means to the maintenance of the moral-ethically defined social order.[5]

Looking into the civil examination system, a critical ladder of success in traditional China, science and technology alone was never an important part of this salient Chinese social-intellectual institution. From the works of traditional Chinese literati, it is not difficult to perceive their frustration towards the unaccommodating mainstream intellectual atmosphere for the pursuit of a pure (or value-freed) technological knowledge. A direct testimony can be found in the preface of Song Ying-Xing's 宋應星 *The Exploitation of the Work of Nature* 天工開物, where it is stated that 'I would advise those brilliant literati, who are longing for their great careers, to throw this book away from their desks, because this book is not going to have any tiny little relevance to the achieving of their scholarly honour, or the pursuit of their official ranks.'[6] Since, as Deng also agrees, the 'natural worldview' (i.e. the fundamental understanding of relations between natural science and human value) of pre-Jesuit China and post-Renaissance Europe were almost mutually exclusive, to leave unexplored the issues of value judgement and the 'cultural logic'[7] of why China had undertaken a different path in scientific investigation, is itself swinging toward the dominant modern European benchmark.

On the so-called 'software' (scientific principles and knowledge) of the steam engine, the pre-Jesuit Chinese science was characterized by abstract reasoning, intuitive deduction, physical observation and direct analogy. As for areas that went beyond intuitive deduction, Chinese scientists tended to employ ambiguous and abstract, yet subtle, metaphysical analogies, such as the equilibrium of *yin* and *yang* forces and the five elements, or the function of *qi* and *li* 理 etc. for conceptual explanation.[8] Scientific theorizations based on these balanced natural forces, as Deng suggests, are not particularly accommodating to the utilization of machinery as motivating powers, whilst adhering to the integrative view of a moral, ethical and natural world sits uneasily with the notion of a vacuum. However, although there was no clear concept of energy conversion and a man-made engine in premodern China, usages of gun powder and applications of air explosion in launching firing arms, rockets and cannons certainly challenge the remark that China never escaped from the 'traditional power trap' of crude natural forces and muscle.[9]

As for the 'hardware' or the machine itself: despite defects, little doubt is shed that in China around 1200, there existed almost all the key technological components of a reciprocating steam engine. In *Wang Zhen's Treatise on Agriculture* 王禎農書,[10] the description and illustration of a 'water-powered metallurgical furnace' (Deng: Figure 16) clearly showed the combination of crank, connecting rod and piston-rod with the animate and water powers, as well as the application of a standard inter-conversion of rotary to rectilinear reciprocating motion in heavy-duty machines.[11] From the 'box-bellows' (Deng: Figure 15), it revealed the artisans' acknowledgement of knowledge of atmospheric pressure, the appliances of double-action valves, pistons, cylinders and air-pumping principles, which worked to suck and expel air.[12] Moreover, in Song's *Work of Nature*, techniques of 'mercury refinement' demonstrated the knowledge of vaporizing metal or liquid mercury into gas, and the usage of a heating boiler: a metal pipe, which connected to an airtight condenser.[13] Finally, in the brine drawing and processing (Deng: Figures 6 and 7), with slight modification, the crude bamboo technology could have feasibly been converted into a huge suction-lift water-pumping machine.

We are not indeed arguing that, given sufficient time, China would eventually have invented a steam engine of its own, as most people would agree that to have most technological component parts is one thing, and to manipulate these fragmented technological components into a combined steam engine is something quite other. The creation of a steam engine involved not only intimate theoretical (or scientific) and technological interactivity, but also a complex socio-cultural-institutional connectivity. The significance of the availability of these component technologies and practical knowledge in sixteenth–eighteenth-century China, however, is that they undoubtedly provided salient conditions for China to react positively to Western technology. In other words, in strictly technological terms, China was fully capable of adopting and diffusing the technology of the steam engine in the eighteenth century if the mainstream Chinese literati had ever really wanted to. What seemed to be inhibiting this Western invention and its diffusion in early Qing China were thus more likely to be non-technological factors. Examining the cultural-historical context we nevertheless come to the same conclusion as Deng: that 'the ultimate reason has to be found outside the domain of technology itself'.

Notes

1. Kent G. Deng, 'Why the Chinese Failed to Develop a Steam Engine', *History of Technology*, 2004, this current volume.

2. Representative figures include scholars such as Zhou Dun-Yi 周敦頤 (1017–73), Cheng Yi 程頤 (1032–85), Cheng Ying 程穎 (1033–1107), and Zhu Xi 朱熹 (1130–1200).

3. Wen Tian-Xiang 文天祥, 'Ode of the Qi of Righteousness 正氣歌, (1238–82), collected in Jian Ting-Xi 蔣廷錫 and Chen Meng-Lei 陳夢雷 (eds.), *The Complete Classics Collection of Ancient China 古今圖書集成* (Taipei, Electronic Database of UDB, 2003. Website: libdb.wtuc.edu.tw/chinesebook/home/index.asp).

4. Du Feng-Xian 杜奉賢, *The Developmental Theory of Chinese History: A Comparison between Marx and Weber's Theory on China 中國歷史發展理論: 比較 馬克思與偉伯的中國論* (Taipei, 正中書局, 1997), 134.

5. Gang Deng, *Development Versus Stagnation: Technological Continuity and Agricultural Progress in Pre-modern China*, (Westport and London, 1993), 28.

6. Song Ying-Xing 宋應星, *The Exploitation of the Work of Nature 天工開物*,(Taipei, 金楓出版, Vol. I, (reprinted 1986), preface, 13.

7. Here 'cultural logic' is understood as the disparate mode of thinking and response in the meaning-making system of a culture, which interprets the significance of life and the function of scientific knowledge within its specific value context. Chinese culture has its own trajectory. Such a trajectory of Chinese culture (although it was by no means determined) was from time to time oriented by the 'moral-ethical-commonsensical'-based rationality that subjected the material and technological progress and practical knowledge to the order of virtue. This cultural logic tends to shift the focus of Chinese intellectuals to moral and ethical aspects of day-to-day life, even in their pursuits of scientific and technological knowledge. Further discussion of the 'moral-ethical-commonsensical'-based rationality can be found in Jin Guan-Tao 金觀濤 and Liu Qing-Feng 劉青峰, *The Origins of Modern Chinese Thought – The Evolution of Chinese Political Culture from the Perspective of Ultrastable Structure (Vol. I) 中國現代思想的起源—超穩定結構與中國政治文化的 演變* (Hong Kong, 2000), chs. 1–3.

8. Jin Guan-Tao 金觀濤 and Liu Qing-Feng 劉青峰, *Prosperity and Crisis 興盛與危機* (Taipei, 風雲時代出版公司, 1994), 439; also Liang Shu-Ming 梁漱溟, *The Essence of Chinese Culture 中國文化要義* (Taipei, 里仁書局, first edition, 1982), chs. 5–7.

9. Kent G. Deng, *op. cit.* (2003).

10. Wang Zhen 王禎, *Wang Zhen's Treatise of Agriculture 王禎農書* (Taipei, 藝文, reprinted 1971).

11. Joseph Needham *et al.*, *Science and Civilisation in China* (Cambridge, 1965) vol. IV, part
II, 374; also Clive Ponting, *World History. A New Perspective*, (London, 2000), 367.
12. Joseph Needham, *Clerks and Craftsmen in China and the West*, (Cambridge, 1970), 155–6.
13. Song Ying-Hsing 宋應星, *op. cit.* (6) (Taipei, 金楓出版 Vol. II, reprinted 1986), 276, 291.

References

Gang Deng, *Development Versus Stagnation: Technological Continuity and Agricultural Progress in Pre-modern China* (Westport and London, 1993).

G. Kent Deng, 'Why Chinese Failed to Develop a Steam Engine', *History of Technology*, this current volume (2004).

Jin Guan-Tao 金觀濤 and Liu Qing-Feng 劉青峰, *Prosperity and Crisis* 興盛 與危機 (Taipei, 風雲時代出版公司, 1994).

___ *The Origins of Modern Chinese Thought – The Evolution of Chinese Political Culture from the Perspective of Ultrastable Structure (Vol. I)* 中國現代思想的起源—超穩定結構. 與中國政治文化的演變 (Hong Kong, 2000). Title translated by the authors.

Liang Shu-Ming 梁漱溟, *The Essence of Chinese Culture* 中國文化要義 (Taipei: 里仁書局, 1982), first edition.

Joseph Needham, *Science and Civilisation in China* (Cambridge, 1965), vol. IV, part II.

— *Clerks and Craftsmen in China and the West* (Cambridge, 1970).

Clive Ponting, *World History. A New Perspective* (London, 2000).

Song Ying-Xing 宋應星, (1637), *The Exploitation of the Work of Nature* 天工開物 (Taipei: 金楓出版, vol. I, II, reprinted 1986).

Du Feng-Xian 杜奉賢, *The Developmental Theory of Chinese History: A Comparison between Marx and Weber's Theory on China* 中國歷史發展 理論: 比較馬克思與偉伯的中Ⅰ 與偉 Taipei: 正中書局, 1997).

Wen Tian-Xiang 文天祥. (1238–82), 'Ode of the Qi of Righteousness   正氣歌 Collected in Jian Ting-Xi 蔣廷錫 and Chen Meng-Lei 陳夢雷 (eds.), *The Complete Classics Collection of Ancient China* 古今圖書集成 *(Taipei, Electronic Database of UDB, 2003. Website: libdb.wtuc.edu.tw/chinesebook/home/index.asp)*.

Wang Zhen 王禎 (1304), *Wang Chen Treatise of Agriculture* 王禎農書 (Taipei, 藝文, reprinted 1971).

The Development of the Steam Engine from Watt to Stephenson

RICHARD L. HILLS

THE IMPACT OF THE STEAM ENGINE

Steam!, all powerful steam! ... It is a subject of great importance to all classes of society and to almost every country. Every one can observe its effects in the ponderous and powerful machines call '*Steam* Engines', now universally used in this and almost every civilized country for the various purposes of driving machinery in manufactories, supplying cities and towns with water, impelling along railroads carriages containing the most heavy and costly merchandize, cattle and passengers, with a velocity which may be said to outstrip the wind, and, despite the tides, propelling vessels on rivers and seas, at a rapid and a certain rate, heretofore dependent upon the winds, which are mutable even to a proverb; and in various other inventions which more or less affect the interests and influence the actions of all mankind.

Whilst those unacquainted with the application of steam can only observe and wonder at its effects; and those who understand both its nature and effects are only intent upon perfecting their theories or improving their machines, it furnishes to the political economist matter of enquiry and reflection of the most interesting and paramount importance, tending, as the power of steam undoubtedly does, from its potent action and extensive application, to produce to an unprecedented extent almost every manufactured article of home consumption and foreign commerce.[1]

So wrote Alderson in 1834. The steam engine had been changed dramatically from the days in late 1763 when James Watt was experimenting with the model Newcomen atmospheric steam engine belonging to the University of Glasgow. At that time, the steam engine was restricted virtually to pumping water out of mines. The situation was completely changed by 1834, when, in manufacturing districts, the largest steam engines probably developed more horsepower than the largest waterwheels. Figures compiled by J. Kanefsky (Table 1) show that, by 1830, the total horsepowers of waterwheels and steam engines in Britain were equal, with that from wind power having been left well behind.

Table 1   *Wind, water and steam power, 1760–1830*

Date	Wind h.p.	%	Water h.p.	%	Steam h.p.	%	Total h.p.
1760	10,000	11.8	70,000	82.3	5,000	5.9	85,000
1800	15,000	8.8	120,000	70.6	35,000	20.6	170,000
1830	20,000	5.7	165,000	47.1	165,000	47.1	350,000[2]

But, as Alderson pointed out, it was more the application of the power of steam to transport which captured the imagination of the general public. At a public meeting held at the Freemason's Hall on 18 June 1824 to consider erecting a monument to James Watt, Humphry Davy said about steam navigation:

> Not only have new arts and new resources been provided for civilised man by these grand results, but even the elements have to a certain extent been subdued, and made subservient to his uses; and, by a kind of philosophical magic, the ship moves rapidly on the calm ocean, makes way against the most powerful stream, and secures her course, and reaches her destination, even though opposed by tide and storm.[3]

While the effect of steam at sea securing a safe, reliable passage was dramatic enough, what steam achieved on the railway was even more spectacular. In 1825, Charles Sylvester pointed out that a horse, when standing still, could prevent a weight of 169 lbs hung over a pulley from falling; at 2 m.p.h. this force had declined to a pull of 100 lbs which rapidly decreased with rising speed so that 'at the speed of 13 miles he is not able to exert any power'.[4] The potential of the steam railway locomotive for power and speed was anticipated for a long time. Considering the primitive state of the steam railway locomotive in 1821, Edward Pease's vision for railways was certainly remarkably prophetic when he wrote to Thomas Richardson,

> don't be surprised if I should tell thee there seems to us after careful examination no difficulty of laying a railroad from London & to Edinburgh on which waggons would travel & take the mail at the rate of 20 miles per hour … We went along a road upon one … These Engines conveying about 50 tons at a rate of 7 or 8 miles per hour, & if the same power had been applied to speed which was applied to drag the waggons we should have gone 50 mile per hour.[5]

However, it was not until the Rainhill Trials organized by the Liverpool & Manchester Railway in 1829 that George and Robert Stephenson's *Rocket* showed a decisive superiority over the horse by travelling at $29\frac{1}{2}$ m.p.h., presumably a world speed record.[6] The pace of development was extraordinary so that Nicholas Wood could point out in 1838,

When the Liverpool and Manchester railway was established, it was made one of the stipulations, at the celebrated contest [Rainhill Trials], that none of the engines should weigh more than five tons, and that the rate of travelling should not be less than ten miles an hour. We now find, the very engine for which the premium was obtained [*Rocket*] discarded as useless, and doomed to drag coal along a private railway, and engines employed upon that railway weighing upwards of twelve tons, while the public are complaining when the rate of travelling is less that twenty miles an hour.[7]

THE INFLUENCE OF JAMES WATT

Perhaps the contemporaries of the Stephensons ascribed too much to the genius of Watt in laying the foundations for this dramatic development of steam engines in so many different spheres. For example, William Huskisson claimed,

Gentlemen, whether ... we look as men to the benefits which Mr. Watt's inventions have imparted, and are still imparting, to the whole race of man; or whether, as members of that great and powerful community of which he was a member, we confine ourselves to contemplate the special benefits which he conferred upon this country, – his great discoveries must stand equally entitled to our highest admiration. As Englishmen, we cannot behold the results produced by his genius, without a lively sense of joy that we belong to the same country to which he belonged; and without an individual feeling of gratitude that he lived at a time which allows us all to participate in the benefits which he was the selected instrument, under Providence, of introducing among us.[8]

Watt's most notable achievements were the separate condenser, together with using steam instead of the atmosphere to propel the piston, patented in 1769, which dramatically improved the efficiency.[9] In 1782, he patented the double-acting engine which gave a power stroke in both directions, thus almost doubling the power output as well as adding another step in the direction of his rotative engine.[10] This patent also included cutting off the steam early during the stroke and so allowing it to expand with a consequent fall in pressure, something developed later to improve engine efficiency. His final engine patent, in 1784, covered the parallel motion which proved to be the key to a successful rotative beam engine.[11] The importance of the development of the rotative engine to the firm of Boulton & Watt may be seen from early production figures of its engines. Between 1775 and 1800, there were ordered from them a total of 496 engines, of which 38 per cent were reciprocating types with 164 pumping and 24 blowing engines. The remaining 308, or 62 per cent, were rotative, built between 1784 and 1800.[12]

There were, of course, many others besides Watt who contributed to the further development of the steam engine, and their inventions are summarized briefly here. Richard Trevithick pioneered the introduction of high-pressure non-condensing types. In 1800, he sent a portable one to Wheal Hope, followed by a larger one at Tredegar Iron Works in 1801 which was being run at pressures from 50 to 100 p.s.i. This was the year that he tried out his road locomotive at Camborne and then in 1804 he introduced a railway locomotive at Penydarren.[13] Higher steam pressures allowed a more successful use of compound engines which Arthur Woolf began exploring around 1804.[14] Freedom from the restrictions imposed by Watt's patents allowed the Cornish engineers to improve the performance of their engines during the first decades of the nineteenth century through using higher pressure steam and greater expansion with early cut off. A contributory factor in this was Trevithick's Cornish boiler which appeared around 1812. This was the cylindrical type with a single fire tube made from wrought-iron plates that proved more reliable than the cast iron advocated by Woolf. While these advances had little impact on those rotative engines used for example in textile mills, the Stephensons did construct stationary beam engines with higher pressures for haulage of railway trains up inclines in the later 1820s.[15]

Yet, when we examine these developments, we find that few experiments were carried out to determine the best form that these inventions should take and that there seem to have been few attempts to expand a theoretical understanding of the steam engine, with the exception of Watt and John Smeaton. Watt's flash of inspiration in the spring of 1765 which resulted in the separate condenser was based on a series of experiments, first on the model Newcomen engine belonging to the University of Glasgow and then on larger ones of his own construction. Joseph Black's explanation of the principles of latent heat added the final piece on the jigsaw of what was happening to the steam as it was generated in the boiler and afterwards at condensation. Watt determined this through using three scientific instruments: the thermometer, the barometer for measuring the atmospheric pressure as well as in the form of a manometer for measuring pressures inside the cylinder, and the balance. His separate condenser could not have been achieved without the discovery by Evangelista Torricelli of the barometer and Gabriel Fahrenheit's accurately graduated thermometers. And, of course, he could not have worked his separate condenser without Robert Boyle's invention of the air pump. Watt's crucial apparatus was based on the equipment pioneered during the seventeenth-century Scientific Revolution. While he recognized the significance of the part played by heat through measuring the temperature of the steam entering the cylinder and the subsequent condensate, he never seems to have enquired into the nature of heat itself, being content to follow the doctrine of phlogiston or caloric, even if he had his doubts.

We may contrast Watt's breakthrough with his separate condenser with Smeaton's trials on his Newcomen engine at Austhorpe. The engine Smeaton had erected earlier for the New River Company in London consumed more coals than he had anticipated. So he carried out a series of trials on his smaller engine, altering one part at a time to arrive at the optimum performance. With this knowledge and experience, he improved the economy of the engines he erected later. He tested the original Newcomen engine at Long Benton Colliery, near Newcastle upon Tyne, and found it was capable of a duty of 5,044,158 lbs. While he increased the duty in his replacement engine to 9,636,660 lbs, he never matched the performance of Watt's engines.

On the other hand, when we examine Watt's attempts to bring his separate condenser to a commercial proposition, we discover that he was far from systematic, constantly changing his designs and constructing totally different types. He must have been a sore trial to John Roebuck, but Black, Roebuck and finally Matthew Boulton continued to support him in spite of many failures. Was this because Watt claimed that he had invented the perfect engine? Soon after his flash of inspiration, Watt wrote to Roebuck on 23 August 1765, 'I have set about a larger and more perfect model, having now little doubt of its performing to satisfaction'.[16] A little later, he told James Lind, 'In short, I expect almost totally to prevent waste of steam, and consequently to bring the machine to its *ultimatum*'.[17] This was a time when people sought perfection in many spheres, for example classical architecture. Watt would have known about two examples in the world of technology. It was not long since John Robison had returned from Jamaica where he had been the official observer at the trial of one of Harrison's chronometers and no doubt would have told Watt about the near perfection of that timepiece. Another example was the performance of waterwheels where their efficiency was compared with a conceptual perfect one which would raise back to the same height as much water as had driven it. This concept had been postulated by Antoine Parent and others on the Continent and it was also used by Smeaton in his account of his improvements to waterwheel design in his paper to the Royal Society in 1759.

Watt was certainly aware of Smeaton's work during his surveying career and also earlier because Robison obtained many books for him from Glasgow University library. Robison said that Watt,

> was confined to his business; I was more at large, and going about the College. I ransacked the libraries for every book that he wanted; and every quotation that he met with made him impatient till he got at the original. I saw every book that he got by any other channel besides the public libraries so I may safely say that I knew the whole extent of his reading.[18]

If only we knew today the extent of Watt's reading, we might have a better idea of the origins of some of his other inventions, such as the sun-and-planet gear and parallel motion.

If we look at the different ways of obtaining rotary motion from a reciprocating, rectilinear one in Watt's 1781 patent,[19] which included his famous sun-and-planet device, we can see that these were the sort of mechanisms depicted in the books of pictures of mechanical devices like those of Ramelli or Zonca printed around the end of the sixteenth century and into the seventeenth. Among Watt's models in the collections of the Science Museum are further examples, one of which is a ladder rack on the end of a connecting rod meshing with a gearwheel, certainly a seventeenth-century mechanistic approach. Another example to be found in Ramelli is a form of parallel motion driving marble saws.[20] Yet, in this case, the origin for Watt's application to the steam engine may have been derived from his perspective-drawing machine which he improved from an example which Lind sent him having seen it in India.[21]

In these books, we also find representations of crank-operated reciprocating pumps, generally driven by hand. We are well aware of Watt's familiarity with the crank through his fury when one of his patternmakers, Dick Cartwright, gave away the secret of Watt's schemes to his rivals one night in the pub.[22] There are two interesting points behind this event. One is that these early pictures do not show a crank being operated by a rectilinear motion and a connecting rod. Then among the early steam engineers, there seems to have been no realization that the piston of an atmospheric engine, which had a variable stroke, could be linked to the fixed length of stroke controlled by a crank. Instances can be quoted of Triewald, Wasborough and Symington, all trying various arrangements of racks, ratchets, pawls and so on.

One of the few early instances of a crank operated through a connecting rod was on the humble domestic spinning wheel. Various governmental agencies in Scotland, such as the Board of the Forfeited Estates, had been giving grants during the middle years of the eighteenth century for supplying such wheels to spinning schools to improve the quality of spun linen.[23] Generally on these wheels, the foot pedal only pulled the connecting rod down, with the momentum bringing the rod up again. However, on some wheels for spinning fine linen, the pedal was pivoted differently, so that the connecting rod both pushed and pulled, giving a smoother motion. Watt must have known about these spinning schools. Such an arrangement had been used centuries earlier on Chinese silk-throwing and reeling machines.[24] Similar ones were employed in the silk industry in London and elsewhere in Britain. Could Watt have seen any during his visits to London? He was after all intensely interested in many mechanical devices, including other machines in the textile industry.

Another intriguing link back to early Chinese technology can be found in the double-acting engine. Joseph Needham has pointed out the use of double-acting bellows and flame throwers by the Chinese at least as early as 1637.[25] Even if Watt knew nothing about these Far Eastern machines, one of his first steam engines with a separate condenser was supplied to John Wilkinson in 1776 for driving a single-acting blast-furnace blowing cylinder.[26] In the following year, Wilkinson was supplied with a second blowing engine at Wilson House in Lancashire. John Farey described the blowing cylinder as, 'closed at both ends, and provided with suitable valves, blew out the air, both in ascending and descending . . . The air from this double blowing cylinder was received into another regulating cylinder'.[27] About five years later, Watt patented his double-acting steam engine. Here is an intriguing link in the origin of the double-acting steam engine which does not seem to have been noticed previously.

THE HIGH-PRESSURE ENGINE

While Watt considered that he had invented the perfect steam engine, it was soon shown that he had not developed the most efficient. In 1769, Smeaton tested fifteen Newcomen engines in the Newcastle upon Tyne area and found that the worst returned a duty of 3,220,000 lbs and the best one of 7,440,000 lbs. In 1778–9, Watt claimed that one of his engines should do 23,400,000 lbs, but it is doubtful whether any of his early engines ever reached this figure, less than 19,000,000 lbs being more realistic.[28] Extensive trials were carried out on the Watt engine at Wheal Harland in 1798 when a duty of 27,500,000 lbs was achieved. Watt stated at the time that it was so perfect that further improvement could not be expected.[29] It is at this point that the theory and practice of the steam engine really began to move into completely new territory. Now Watt advocated low-pressure steam. He was no stranger to using steam at higher pressures, for example with his steam wheel, but one reason for retaining low pressure was a boiler explosion at Poldory Mine in January 1784 when there were several fatalities.[30] Another may have been his theory of Watt's Law which stated that the sum of latent and sensible heats always remained the same, which showed no advantages for higher pressures.

The early nineteenth-century Cornish engineers led by Trevithick had no such qualms. Trevithick installed several column-of-water or water-pressure engines in mines there. In essence, these were similar to steam beam engines but driven by the weight of a column of water, yielding higher pressures and hence the need for smaller cylinders and pistons than equivalent steam engines. Such engines had a history which can be traced back to the seventeenth century but the characteristic form only emerged in the middle of the eighteenth in Schemnitz (Slovakia), in France and a few years later in England.[31] In such engines, since water is a virtually incompressible fluid, there can be no early cut off with consequent expansion and a fall in the pressure of the fluid in the cylinder. The head of water of one of Trevithick's engines at Wheal Druid was 204 ft, or about 600 p.s.i.[32]

Trevithick's early high-pressure steam engines were nicknamed 'puffers' because they exhausted directly into the atmosphere being non-condensing with no cut off or expansion. This is confirmed by an account of the whim or winding engine at Cook's Kitchen Mine because a Mr Hunter recollected 'the valley engine, because she was a puffer, you could hear her for miles'.[33] Another engine supplied to Herland Mine in 1816 was also non-condensing from an account given by James Banfield:

> When a young man, living on a farm at Gurlyn, I was sent to Gwinear to bring home six or seven bullocks. Herland Mine was not much out of my way, so I drove the bullocks across Herland Common toward the engine-house. Just as the bullocks came near

the engine-house the engine was put to work. The steam roared like thunder through an underground pipe about 50 feet long, and then went off like a gun every stroke of the engine. The bullocks galloped off, some one way and some another. I went into the engine-house.[34]

That Trevithick at first did not use steam expansively has been confirmed recently through the construction of a replica of his 1802 patent road carriage which gives a sharp blast of steam at the exhaust.

Watt had recognized the economic advantages of an early cut off in his 1782 patent.[35] If the valve admitting steam to the cylinder were shut part way through the stroke, the steam would expand while its pressure would fall. He followed Boyle's Law to determine the pressure drop and drew a diagram in his patent. He found that,

> It appears that only one-fourth of the steam necessary to fill the whole cylinder is employed, and that the effect produced is equal to more than one-half of the effect which would have been produced by one whole cylinder full of steam, if it had been admitted to enter freely above the piston during the whole length of its descent.[36]

While he had halved the power, he had quartered the fuel consumption. Why this happened was not understood properly at that time.

Trevithick was quick to discover the same phenomenon on an engine driving a rolling mill at Penydarren in 1804. 'When the cylinder was full of steam the rollers could not stop it; and as coal is not an object here, Mr. Homfray wished the engine might be worked to its full power. The saving of coal would be very great by working expansively'.[37] But of course coal was expensive in Cornwall where soon steam at higher pressures, worked expansively, was quickly taken up. In 1811, Captain John Davey of Gwinear started to report the duty of his engines at Wheal Alfred which were considered to be the best in the county and returned about 20,000,000 lbs duty. Captain Joel Lean took over the reporting and, in 1814, three engines at Dolcoath returned duties of 21,445,912 lbs, 26,756,063 lbs and 32,027,842 lbs.[38]

About this time, Woolf returned to Cornwall where he started erecting his double-cylinder compound engines. One at Wheal Abraham was tested in October 1814 and returned a duty of 34,000,000 lbs. After some faults had been rectified, its highest duty of 55,900,000 lbs was recorded in May 1816.[39] While this performance remained exceptional, constant small improvements over the years continued to raise the duty figures. In 1835, the hitherto steady increase in duty was shattered by the starting of Austin's 80 ins engine at Fowey Consols which, during that July, averaged 90,000,000 lbs. It had a public trial in October when an incredible 125,000,000 lbs duty was recorded.[40]

At this period, the steam engine was considered primarily as a machine driven by the pressure of the steam rather than as one which converted heat into mechanical energy. It could be easily compared with a waterwheel, where the pressure corresponded with the height of the fall of the water, and of course the volumes of steam or water were then comparable. The expansion of the steam through its cut off might be compared with that part of the distance through which the water had fallen. If this were true, then it would make little difference whether a large volume of water or steam falling a short distance or a small volume falling a greater distance were used to achieve the same power output. In the case of steam, there was greater safety with lower pressures. But this left unanswered the apparent greater efficiency of steam at higher pressures.

We can see the concept of the steam engine as a pressure engine reflected in the text of Woolf's patent for his compound engine of 1804:

> By small additions of temperature, an expansive power may be given to steam to enable it to expand to fifty, sixty ... three hundred, or more, times its volume without any limitation but what is imposed by the frangible nature of every materials of which boilers and the other parts of steam engines have been or can be made.[41]

The expansive principle seemed to have potential to yield almost boundless power. This is reflected in Farey's contribution to Rees's *Cyclopaedia*, published in 1816:

> We have been thus full upon this subject, because the gaining more power by the expansion of air or steam acting in double cylinders, has been a favourite idea with many, and there are no less than five different patents for it, but several of these have been upon mistaken notions ...
>
> The advantage of the expansive principle arises wholly from a peculiar property of steam, by which, when suffered to expand itself to fill a greater space, it decreases in pressure or elastic force by a certain law, which is not fully laid down; that is the relationship between its expansive force and the space which it occupies is not clearly decided.[42]

Farey himself was one of those with 'mistaken notions' because the science of thermodynamics did not exist until many years later. The role of heat being transformed into energy accompanied by a fall in temperature was not understood. The caloric theory in which heat was some form of subtle fluid reigned supreme and so the mystery of the reason for the greater economy of the high-pressure expansive engine remained unsolved. Perhaps Watt's concept of his perfect engine remained dominant and others were afraid to challenge it. Certainly Thomas Tredgold thought this.

> The idea that Watt had done everything possible to be done reflecting the power of steam had stopped inquiry among men of science, and left the manufacturers and capitalists of the country, who were wishful to encourage improvement, to be guided by vain and ignorant projectors or ruined by pretending knavery.[43]

THE STEAM ENGINE APPLIED TO TRANSPORT

The period covered by this paper saw the steam engine applied to transport on both water and land. In both applications, it was a case of slow adaptation of existing types to suit the special circumstances without any development of new theories. For ships, safety from boiler explosions must have been a prime consideration so here the low-pressure steam engines with condensers reigned supreme at the expense of enormous engines and boilers. On the other hand, engines employed either on rail or road needed to be compact but powerful so that higher pressures were essential to supply small non-condensing engines. Development of the mechanisms in both cases was empirical, being based mainly on experience gained through the operation of earlier types.

THE STEAMBOAT

Between the first serious experiments made by Patrick Miller and William Symington in 1788 and what was probably the first practical steamboat, the *Charlotte Dundas*, built in 1802, there were many failures because a successful type of engine had not been evolved. A reliable engine, with a reasonable power-to-weight ratio had first to be developed and Watt held the key to this until 1800 with his patent for the separate condenser as well as the double-acting cylinder. It is interesting to note that nearly all the early experiments were made on lakes or canals where the boat did not have to battle against tides or river currents. Possibly the steamboat might have appeared sooner if Britain had had a large chain of lakes where a boat could have manoeuvred freely without the fear of washing away the banks or being swept away by the current.

Once a sufficiently powerful engine of suitable design had been evolved, then the steamboat could master the currents and move out into the rivers and tidal estuaries. Robert Fulton ordered one of Boulton & Watt's double-acting bell-crank engines to power his *Clermont* which made its first trip up the Hudson River in the USA on 17 August 1807, marking the inauguration of steam navigation as a regular means of transport.[44] As far as Britain was concerned, Scotland was the real birthplace of the steamboat. The *Comet* was soon joined by another ship, the *Elizabeth*, in 1812 to work between Glasgow, Greenock and Helensburgh, and, by 1818, there were 18 steam vessels regularly employed on the Clyde, two on the Tay, two at Dundee and six on the Forth. Already in 1814, the *Marjory* had ventured south across the open sea and was employed on the Thames while, in 1815, the *Elizabeth* left the Clyde to become the first Mersey ferry, travelling between Liverpool and Runcorn. In 1817, James Watt junior fitted engines and paddles to the *Caledonian* and made the voyage from London to Margate that August.[45]

In 1819, the steamboat really began to conquer the high seas since, in that year, the *Waterloo* of 200 tons and 60 h.p., the largest steamer of her year, inaugurated the Liverpool–Belfast route while, in the same year, the Liverpool–Clyde service was started and the *Rob Roy* maintained the

Greenock–Belfast route. In 1822, the *James Watt* was regularly plying between London and Leith. In the 1820s, steamboat development was rapid and it is fitting to conclude this brief review with the first trans-Atlantic service started by the *Sirius*, 320 h.p., and the *Great Western*, 750 h.p., in 1838. These voyages showed that the steamship could conquer the oceans as well as the more sheltered seas or coastal waters.

Yet, while the progress of the steamboat was rapid, after an intial promising start the technological development of the nautical steam engine slowed down, so that engines in these later ships had virtually advanced only in size compared with those of 1820 or even 1812. The first boat designed and built by Symington had twin hulls with a pair of open-topped atmospheric cylinders, 4 ins diameter by about 18 ins stroke. The pistons were connected by chains to a drum which turned in opposite directions alternately. On the paddlewheel shaft were loose pulleys with pawls and ratchets. Chains from the drum turned the loose pulleys in alternate directions but, through the ratchet gear, the paddlewheel turned continuously in one direction only. Although the vessel on her trials on Dalwinston Lake near Dumfries on 14 October 1788 achieved the speed of 5 m.p.h.,[46] such a mechanical contraption was too complicated to have any future.

First, the drive to the paddlewheels had to be simplified and, in 1798–9, an experimental boat built at Worsley near Manchester for trials on the Bridgewater Canal had a single-cylinder open-topped atmospheric engine connected to a large flywheel. Shafts and bevel gears turned the paddleshaft. John Smith of St Helens in Lancashire and Symington on the *Charlotte Dundas* in 1802 put a crank on the paddleshaft and drove it either through a beam and connecting rod or directly by a connecting rod. Then the engine itself had to be improved. The atmospheric steam engine had a power stroke only as the piston went into the cylinder and some other force had to pull the piston back or, more usually, up again. Two cylinders working alternately would have solved the problem, but this meant a large increase in useless weight. Also the effective pressure on the piston was only 7 or 8 p.s.i. It is significant that all those boats which used atmospheric engines were failures. The trials on Dalwinston Lake were soon stopped. No more was heard of John Smith after his voyage on the Bridgewater Canal from Runcorn to Manchester. The boat built at Worsley as a barge tug hauled its train no more quickly than a horse and it was feared that the wash would damage the banks. In fact this boat was so unpopular that it was nicknamed *Bonaparte*.[47]

The first successful steamboats were pioneered in the surge of new steam engine inventions which followed the expiry of Watt's patent in 1800. New valve gears and different types of valves, different layouts with side levers or bell cranks instead of the main beam, table engines, etc., were all developed in the first years of the nineteenth century, and most of them found their way at sometime into steamboat design. Symington in the *Charlotte Dundas* rather exceptionally for Britain used a single horizontal cylinder, double-acting, 22 ins diameter by 4 ft stroke of 10 nominal horse power driving directly by a connecting rod to the crank on the end of the paddlewheel shaft at the stern. The reason for the adoption of this layout was the need to fit the boat into the Forth & Clyde Canal. She was built to haul barges on that canal and, on her tests in March 1802, hauled two loaded vessels, each of 70 tons burden, for 19.5 miles in 6 hours against a strong headwind. Although the tests were successful, the canal company decided that any advantages from steam haulage would be more than outweighed by the damage done to the banks.[48]

The type of steamboat which was to become the most popular in this country had a pair of paddlewheels situated one on either side of a single hull, roughly in the middle. In the days when horizontal cylinders were regarded with disfavour because they could not be oiled properly and wore badly, and long connecting rods were thought to be essential, the design of an engine to fit into a boat presented peculiar problems. What evolved as the common solution was the inevitable compromise. The weight had to be kept as low as possible to give the maximum stability, so the cylinder was placed vertically near the bottom of the boat. The rocking beam, or more usually a pair of beams, was set beside the bottom of the cylinder and connected to the crosshead and piston rod by long links reaching down instead of up. The other end of the beam drove the crank on the paddleshaft by a connecting rod reaching upwards.

This side-lever engine, at first as a single engine with one cylinder but later as a pair of engines with two cylinders, was used extensively for about 40 years until the early 1850s. Perhaps the next most popular type was the 'vibrating' or 'oscillating' engine introduced by Henry Maudslay in 1827. Here the cylinders were placed in the bottom of the ship below the paddleshaft and drove it directly from the piston rod with no connecting rod. It was a fascinating sight to watch the cylinders gently rocking to and fro as the engine worked. On the steeple engine, the cylinders were again placed in the bottom of the boat under the paddleshaft but were linked to a crosshead high above with a connecting rod hanging down from it to turn the crank.[49]

Space is too limited to describe all the varieties of engines in boats, but, whatever their construction, most of these early ones had boiler pressures of 2 or 3 p.s.i., or possibly 5 p.s.i., with jet condensers drawing cold water from the sea. Typical of this early period was I.K. Brunel's first ship, the *Great Western*, for, although she was the largest vessel afloat, the engine design was quite ordinary. Maudslay built a pair of enormous side-lever engines, $73\frac{1}{2}$ ins bore by 7 ft stroke, which weighed 310 tons. With a boiler pressure of only 5 p.s.i., they produced 750 h.p.[50] Their size and power must have exceeded anything working on land at that time.

One reason for the low pressure was fear of boiler explosions which would have had even worse catastrophic results at sea than on land. High-pressure steam had been used in boats. In 1804 in the USA, Oliver Evans constructed his *Orukter Amphibolos* with a 5 n.h.p. engine worked at 120 p.s.i. Two years later in London, Trevithick had placed one of his high-pressure non-condensing engines on a boat to work a dredger and in 1812 he put another into a paddlesteamer.[51] But such developments were stopped short by the explosion of a boiler in a boat on the River Yare at Norwich in 1817. Woolf and others were examined at the inquiry afterwards and, although it was found that the safety valve had been tied down, the recommendation was made that pressures be kept to $2\frac{1}{2}$ to 6 p.s.i.[52] Understandably people were reluctant to use high pressures after this. Even in the 1830s, the ordinary steam pressure in marine boilers was still only 5 p.s.i. and, in the 1840s, 10 p.s.i.[53] Though this meant loss of cargo capacity and increased fuel consumption through the construction of enormous boilers and engines, the safety of low-pressure steam was preferred as long as there were no scientific reasons or thermodynamic advantages known for using higher pressures.

ROAD TRANSPORT

The steam engine applied to locomotion for either road or rail use presented entirely different problems to the designer than those in ships because here a high power-to-weight ratio was essential. Cugnot's enormous steam carriage could hardly haul itself along, let alone a gun. Murdock, after scaring the local parson out of his wits at Redruth by trying out his model steam carriage in the churchyard one dark night, was dissuaded by Boulton from further experiments.[54] Trevithick did not have much more success either. His first one was tried out at Camborne on Christmas Eve 1801, but was destroyed a few days later while its drivers were celebrating in the local pub. The fire in the boiler was left unattended, the water boiled away, the boiler became red-hot and set fire to the engine and shed.[55] Neither did Trevithick have much better success with the unwieldy road carriage he patented in 1802 and had built in 1803. It made some trial trips in London but succeeded only in tearing down six or seven yards of railings, which ended further test runs.

During the period we are considering, steam locomotion on roads was never introduced successfully. It was not only the high tolls charged by turnpike trusts which caused Sir Charles Dance and Mr Gurney to withdraw their steam coaches between Gloucester and Cheltenham, 'so that the public are thus deprived of the benefit of a cheap, a speedy, and a *safe* conveyance',[56] but the coaches proved to be mechanically unreliable and there was always the fear of boiler explosions. This proved to be only too real when the boiler of Richard Roberts's steam carriage exploded in Oxford Road, Manchester, on 4 April 1834. The pumps for the boiler water supply had not been working properly and the fire was dropped. The boiler was refilled and the fire relit, but damage had been done to the boiler. As a result of the explosion, some men were severely scalded and a shop set on fire. While the *Manchester Guardian* pointed out that such mechanical failures might in time be rectified, Roberts never ran his steam carriage again, no doubt because, by this date, the railway locomotive was showing its superiority.[57] A great deal more development work would have been needed on boilers, brakes, strength of components such as wheels, and a much higher standard of road maintenance before steam road carriages could have succeeded.

RAILWAY LOCOMOTIVES

Trevithick was the first to try adapting the steam locomotive for traction on railways but he had no better success with them than with his ones for the road. His first in 1804 at Penydarren in South Wales was too heavy and broke the cast-iron plate rails.[58] The *Catch-me-who-can*, which Trevithick demonstrated on a circular track at a site near the present Euston Station in London in 1808, overturned when a rail broke. Even with a charge of a shilling a ride, it failed to pay its expenses.[59]

Much had to be done to turn the first single-cylinder springless locomotives into reliable machines capable of hauling a commercial load. This required radical redesign of the layout which evolved piecemeal, based mostly on the operating experience gained over the next 20 years. The complexities of the motion rods linking vertical cylinders to the crankpins on the wheels and the associated valve gear mechanisms had to be simplified. However, it was in boiler design that improvements had to be made to withstand higher pressures, to provide an adequate supply of steam, but above all to reduce the weight.

This was the problem with which George and Robert Stephenson struggled in the 1820s. *Locomotion* at the opening of the Stockton & Darlington Railway in 1825 had a boiler with a single fire tube through it. Here such locomotives hauled the coal trains at speeds little faster than the horses at work on the same line. The *Lancashire Witch*, built for the Bolton & Leigh Railway in 1828, had twin fire tubes, rather like the later Lancashire boilers in textile mills.[60] While it may have performed better, another experiment in the following year with a very peculiar design having a pair of tub boilers on the *Twin Sisters* was not a success.[61]

The solution which would change the whole course of locomotive engineering and rail transport was suggested to Robert Stephenson by Henry Booth, secretary to the Liverpool & Manchester Railway.[62] This was the multi-tubular boiler. It could have been derived from the type of boiler developed by Marc Seguin for a locomotive on the St Etienne Railway. If so, it was one of the few, if not the only, influence on railway locomotive design originating outside Britain in this early period. The multi-tubular boiler was used with outstanding success by Robert Stephenson on *Rocket* at the Rainhill Trials organized by the Liverpool & Manchester Railway in 1829. *Rocket* gave such a brilliant performance in hauling capacity, reliability and speed for the size of the locomotive that there was no longer any doubt that here was the motive power of the future for railways.

The opening of the Liverpool & Manchester Railway on 15 September 1830 inaugurated a new era in the history of transportation and civilization for here was the first true inter-city railway for both goods and passengers. Its popularity among the travelling public exceeded all expectations since it was both quicker and cheaper than the stagecoaches. The capacity of the line proved inadequate so that, in the next few years, it had to be almost completely rebuilt from the rails upwards. This included the locomotive stock, culminating in Robert Stephenson's *Patentee* 2-2-2 design of 1833. Thomas Gray's vision for steam railways in 1822 had proved correct when he prophesied their 'vast superiority in every respect over all the present pitiful methods of conveyance by turnpike-roads, canals and coasting traders'. He wrote,

> No speed with this can fleetest horse compare,
> No weight like this canal or vessel bear;
> As this will commerce every way promote,
> To this let sons of commerce grant their vote.[63]

It was the railway locomotive which really introduced the steam engine to the bulk of the population because they could see the trains huffing and puffing their way through the countryside as the web of lines spread over the land. One such person was James Prescott Joule who, as a young lad, went to a field near Eccles to watch the first trains running on the Liverpool & Manchester Railway. This would have been an experience that endured in his memory and quite possibly influenced the course of his thoughts and future experiments when he investigated the phenomenon of heat, laying the foundation for a true understanding of thermodynamics which would determine the later development of the steam engine.[64] It needed men of science to unravel the mysteries of the way in which steam engines operated. The words of Dionysius Lardner, himself an advocate of the expansive principle, written in 1856 were proved true:

> The general state of physical science at the present moment, the vigour, the activity, and sagacity with which researches in it are prosecuted in every civilised country, the increasing consideration in which scientific men are held, and the personal honours and rewards which begin to be conferred upon them, all justify the expectation that we are on the eve of mechanical discoveries still greater than any which have yet appeared; that the steam-engine itself, with its gigantic powers, will dwindle into insignificance in comparison with the energies of nature which are still to be revealed; and that the day will come when that machine, which is now extending the blessings of civilisation to the most remote skirts of the globe, will cease to have existence except in the page of history.[65]

Notes and References

1. M.A. Alderson, *An Essay on the Nature and Application of Steam* (London, 1834), 1–2.
2. J. Kanefsky, 'The Diffusion of Power Technology in British Industry, 1760–1870', Ph.D. thesis, Exeter 1979, 338.
3. M. Arago, *Historical Eloge of James Watt* (London, 1839), 103.
4. C. Sylvester, *Report on Rail-Roads and Locomotive Engines* (Liverpool, 1825), 14.
5. Quoted by A. Guy, 'North Eastern Locomotive Pioneers, 1805 to 1827: A Reassessment', in A. Guy and J. Rees, (eds.), *Early Railways* (London, 2001), 13; E. Pease to T. Richardson, 10 October 1821.

6. N. Wood, *A Practical Treatise on Rail-Roads* (London, 1838), 314.

7. *Ibid.*, 681.

8. Arago, *op. cit.* (3), 199–200.

9. Patent 913, 5 January 1769.

10. Patent 1,321, 4 July 1782.

11. Patent 1,432, 25 August 1784.

12. See R.L. Hills, *Power from Steam, A Brief History of the Stationary Steam Engine* (Cambridge, 1989), 70; and H.W. Dickinson, *A Short History of the Steam Engine*, (Cambridge, 1938), 88.

13. Hills, *op. cit.* (12), 101–2.

14. T.R. Harris, *Arthur Woolf, 1766–1837, The Cornish Engineer* (Truro, 1966), 32; and Patent 2,772, 1804.

15. Hills, *op. cit.* (12), 142–3.

16. Birmingham Central Library, James Watt Papers 4/32.38, J. Watt to J. Roebuck, 23 August 1765.

17. B.C.L., JWP C1/15, J. Watt to J. Lind, 4 September 1765.

18. J.P. Muirhead, *The Life of James Watt, with Selections from his Correspondence* (London, 1858), 72.

19. Patent 1,306, 25 October 1781.

20. M.T. Gnudi, trans., *The Various and Ingenious Machines of Agostino Ramelli* (New York, 1976).

21. E. Robinson and D. McKie, *Partners in Science, Letters of James Watt and Joseph Black* (London, 1970), 433; J. Watt's Note Book, JWP W/14.

22. S. Smiles, *Lives of the Engineers. The Steam-Engine. Boulton and Watt* (London, 1878), 228.

23. D. Bremner, *The Industries of Scotland, Their Rise, Progress and Present Condition* (Glasgow, 1869, reprint Newton Abbot 1969), 217.

24. J. Needham (ed.), *Science and Civilisation in China*, vol. 5, *Chemistry and Chemical Technology*, part IX; D. Kuhn, *Spinning and Reeling* (Cambridge, 1988), 393–9.

25. *Transactions of the Newcomen Society*, 35: 1962–3; J. Needham, 'The Pre-natal History of the Steam Engine', 15–22.

26. H.W. Dickinson, *John Wilkinson, Ironmaster* (Ulverston, 1914), 328.

27. J. Farey, *A Treatise on the Steam Engine, Historical, Practical and Descriptive* (London, 1827, reprint Newton Abbot 1971), 1: 328.

28. T. Lean, *On the Steam Engines in Cornwall* (London, 1839, reprint Truro 1969), 6.

29. Farey, *op. cit.* (27), 2: 240–1; and Lean, *op. cit.* (28), 7.

30. Cornwall County Record Office, Wilson Letters, DDX/318.1, J. Watt to J. Wilson, 24 January 1784.

31. D.S.L. Cardwell, *From Watt to Clausius, The Rise of Thermodynamics in the Early Industrial Age* (London, 1971), 85.

32. F. Trevithick, *Life of Richard Trevithick, Inventor of the Locomotive Engine* (London, 1872), 1: 78 and 87.

33. *Ibid.*, 1: 92.

34. *Ibid.*, 2: 75.

35. Patent 1,321, 4 July 1782.

36. J.P. Muirhead, *The Origin and Progress of the Mechanical Inventions of James Watt* (London, 1854), 3: 75.

37. Trevithick, *op. cit.* (32), 2: 132.

38. Lean, *op. cit.* (28), 19–20.

39. Harris, *op. cit.* (14), 53.

40. D.B. Barton, *The Cornish Beam Engine* (Truro, 1969), 49–50.

41. Patent 2,772, 1804.

42. A. Rees, (ed.), *The Cyclopaedia; or Universal Dictionary of Arts, Sciences and Literature* (reprint 1972), 5: 114.

43. T. Tredgold, *The Steam Engine, Its Invention and an Investigation of Its Principles for Navigation, Manufactures and Railways* (London, 1838), 41.

44. E.C. Smith, *A Short History of Naval and Marine Engineering* (Cambridge, 1937), 14.

45. B.C.L., JWP C6/4.110, J. Weston to J. Watt, 4 September 1817.

46. H.P. Spratt, *The Birth of the Steamboat* (London, 1958), 49.

47. F. Mullineux, *The Duke of Bridgewater's Canal* (Eccles, 1959), 27.

48. Spratt, *op. cit.* (46), 61–2.

49. J.G. Winton, *Modern Steam Practice* (London, 1888), 347 and 377.

50. I. Brunel, *I.K. Brunel* (London, 1870), 245.

51. Spratt, *op. cit.* (46), 69.

52. J. Guthrie, *A History of Marine Engineering* (London, 1971), 117.

53. Smith, *op. cit.*, (44), 133.

54. L.T.C. Rolt, *James Watt* (London, 1962), 80.

55. H.W. Dickinson and A. Titley, *Richard Trevithick, The Engineer and the Man* (Cambridge, 1934), 48.

56. Alderson, *op. cit.* (1), 80.

57. See R.L. Hills, *Life and Inventions of Richard Roberts, 1789–1864* (Ashbourne, 2002), 171.

58. Dickinson and Titley, *op. cit.* (55), 65.

59. *Ibid.*, 108–11.

60. C.F. Dendy Marshall, *A History of Railway Locomotives down to the end of the year 1831* (London, 1953), 139.

61. *Ibid.*, 142.

62. R.E. Carlson, *The Liverpool & Manchester Railway Project, 1821–1831* (Newton Abbot, 1969), 63.

63. J. Francis, *A History of the English Railway; Its Social Relations and Revelations, 1820–1845*, (London, 1851), 1: 82.

64. D.S.L. Cardwell, *James Joule, A Biography* (Manchester, 1989), 14.

65. D. Lardner, *Steam and Its Uses* (London, 1856), 208.

The Resources of Decisive Technological Change: Reflections on Richard Hills

IAN INKSTER

It took a brilliant Frenchman in the middle of the Steam Decade to recognize that if you were

> to deprive England of her steam engines, you would deprive her of both coal and iron; you would cut off the sources of all her wealth, totally destroy her means of prosperity, and reduce this nation of huge power to insignificance. The destruction of her navy, which she regards as the main source of her strength, would probably be less disastrous.[1]

Carnot was right. But how had Britain arrived at such a position ahead of its immediate industrial rivals? Hills rightly centres on issues of knowledge, craftsmanship and creativity.[2] Were these together the necessary and sufficient resources of the great Watt breakthrough? Watt's own later account, in a letter to James Hutton in December 1795, stressed an established mode of enquiry fusing detailed observation, experiment and suppositions: 'I do not believe even in Mechanicks without experiment to which test I wish to bring all theoretical opinions if possible.'[3] In the following year, Watt elaborated his creative moment during a dispute over patent rights, in an account reported by Muirhouse. This is worth reproducing as it is a precise summary of the general thesis linking creative technological change in Europe to a range of reliable knowledge, skill and experience, and appropriate immediate location.[4]

> A well-made brass model of Newcomen's engine consumed quantities of steam and fuel out of all reasonable or direct proportion with larger engines. He consulted Desaguliers' *Natural Philosophy* and Belidor's *Architecture Hydraulique*, the only books from which he could hope for information. He found that both of them reasoned learnedly, but by no means satisfactorily; and that Desaguliers had committed a very gross arithmetical error in calculating the bulk of steam from the water evaporated in a common steam-engine; which being rectified, it appeared next that his data, or assumed facts, were false. By a simple experiment, W found what was the real bulk of

water converted into steam; and from his friend Dr Black he learned what was the heat absorbed and rendered latent by the conversion of water into steam, which the Doctor then publicly taught, and had done for some years.

We must it seems embrace the inheritance of Newcomen and, in particular, notions of atmospheric pressure and vacuum, we must include the public science environment, and we must add instrumentation – the Boylean air pump, the thermometer, barometer and balance, apart from the general skill in instrumentation that made up the effective milieu.[5] This should be well known, but may I quote how Cardwell quite clearly emphasizes the place of the conception of the vacuum in the famous Glasgow Green 'eureka moment' of May 1765:

> Steam, Watt knew familiarly from his many experiments, as well as knowing it in the abstract, the theoretical way, is an elastic fluid; it will go on expanding into a vacuum. If you make a connection between the working cylinder and the condensing cylinder (hereafter to be called the condenser) and, at the right moment, you inject cold water into the latter, then steam from the cylinder will rush into the condenser as the steam already there is condensed, and the process will continue very rapidly indeed until all, or practically all, the steam in both cylinder and condenser has been condensed. And all the time the cylinder can be kept boiling hot, the condenser cold and the vacuum in both near perfect. All the condensing water and condensed steam will be in the cold condenser, so it cannot boil and ruin the vacuum. This beautiful invention could only have been made, one feels, by a man who had a personal and very realistic appreciation of the behaviour of steam. Great familiarity with, and skill in handling, materials and processes is often the hallmark of the able scientist.[6]

Furthermore, Cardwell goes on to judge that Watt could not have proceeded but empirically, that is without the use of experiments and measurements. In particular, Watt could only know of the cost effectiveness of his more costly engine because he could in advance estimate from experiment with induction that the addition of the separate condenser, the higher engineering standards, the abolition of the crude water seal, and the use of the air pump would combine to increase fuel economy sufficiently to offset all additional costs. In many ways Cardwell is following contemporary opinion: thus Elijah Galloway for identical reasons wrote of Watt 'the great experimentalist', and remarked that the latent heat and related experiments 'were more decisive and useful than any theoretical explanation could possibly be'.[7] Profound science and original genius is the repeated contemporary phrase for Watt, coined first perhaps by Davy but repeated *ad nauseam*. Thus the Hon. Francis Jeffrey on Watt in *The Scotsman* 4 September 1819, 'Perhaps no individual in his age possessed so much and such varied and exact information – had read so much, or remembered what he had read so accurately and well', and

Richard Hills has reported how Robison 'ransacked the libraries for every book that he wanted'. This was compounded in 'his power of digesting and arranging in its proper place all the information he received' – he was never encumbered. Again, the Earl of Liverpool at Freemasons Tavern on 12 June 1824, 'by the sagacity of his mind he was enabled to apply the profoundest principles of science to the practical purposes of life'. Simon Schaffer has emphasized that we must beware such constructed and reconstructed stories of creation.[8] What they imply when they are regarded even so problematically is that the resource base was highly complex, some compound of the individual in his specific site, a site that included resources for 'making up' discovery claims and thus securing property rights.[9]

The bulk of the concern of Richard Hills' paper is with the resources brought to bear on the move from steam as 'interesting novelty' to steam as 'immense motive power reality'. The speed, weight, size and cost features of the atmospheric engine were under real challenge from the expansive power of high-pressure steam from the point when Trevithick and others could depend upon the rapid evolution of strong fire-tube and water-tube boilers and all those incremental innovations designed to increase the area of the heated surface and strengthen the structure. From this time the vacuum and the weight of the atmosphere are no longer providing the working power of the new-style engines. Here it seems that the major lessons from Richard Hills' expert discussion are as follows:

> Higher pressure emerged empirically and in the absence of reliable explanation as to its physical efficiency.

> This seems to be equally true of most transport applications.

> Very significant efficiency improvements (especially when measured in fuel rather than strictly engineering terms) arose from specific, incremental advances, adjustments, changes in materials, etc, as with the Woolf double cylinder compound engine into the 1830s.

The connections of such incrementalism into the 1830s seemed to have been informal, generated by the industrial imperatives and the urban information systems of the time, hardly discussed and decided upon.[10] This reflects our emphasis at this workshop on specified locations and dissemination of useful knowledge. A rough measure of this is provided in the more significant of the steam-engine patents for the years 1801–30, a listing about which I make no claims as to comprehensiveness, although it includes those of Murray, Bramah, Symington, Trevithick, Parkinson, Woolf, Donkin, Hornblower, Lister, Maudslay, Brunton, Rastrick, Stephenson, Losh, Brunel, Perkins, Gurney, Galloway, Neilson and Poole. Of 330 such steam-engine patents, almost all were registered by engineers rather than by craftsmen.[11] Of the least ambiguous or general of the specifications, some 45 per cent were concerned with improvements relating to high pressure (of boilers, pistons), another 14 per cent were concerned with rotary motion, and another 12 per cent specifically with the saving of fuel. Bearing in mind Richard Hills' emphasis on the developments in vessel, carriage and locomotive transport, some 27 per cent are specifically related to propulsion and railways (paddlewheels and cranks, valves, connecting rods, oscillating engines, marine and locomotive boilers and brakes), these mostly crowded into the late 1820s.

The lessons of such an approach from the history of technology for more global perspectives is that the resources of significant technological breakthrough are complex and vary with close circumstance. The resources required for initiation of change may not be those required of its diffusion or its imitation, and these may not be those required for borrowing and imbibing, transferring and applying to new socio-cultural frameworks.

In summary, it seems that the early nineteenth-century diffusion of steam within Britain was not analogous to its expansion under pressure, nor was it osmotic. It was partial and fractive and occurred in specific places as new innovations altered the functions and applications of steam power, such innovations being as commercially important as – and technically related to – the improvement of the actual standing engine itself. This seems to me to be a major lesson of Richard Hills' paper. We might conclude that the resources required for such internal diffusion were similar or even equivalent to those required by a foreign technological system in order to imbibe (effectively transfer-in) the Watt technology. But, we might also maintain that this combination of resources was not sufficient to create original Watt improvements. Are such distinctions worth examining and debating, or are they altogether too reductive for post-modern comforts?

Notes and References

1. Sadi Carnot, *Reflexions on the Motive Power of Fire*, trans. and edit., Robert Fox (Manchester, 1986 (1824)), quote 62.

2. See also Richard Hills, *Power from Steam, A History of the Stationary Steam Engine* (Cambridge, 1989); *idem.*, 'James Watt, Mechanical Engineer', *History of Technology*, 18 (1996); *idem.*, *James Watt, vol. 1: His Time in Scotland 1736–1774* (Ashbourne, 2002); *idem.*, *Life and Inventions of Richard Roberts 1789–1864*, (Ashbourne, 2002).

3. Watt to Hutton, December 1795, in J. Jones, H.S. Torrens and E. Robinson, 'The Correspondence between James Hutton and James Watt', *Annals of Science*, 1995, 52: 357–83, quote 380.

4. James Patrick Muirhead, *The Life of James Watt* (London, 1858), 83–91.

5. See for instance Joel Mokyr, *The Lever of Riches* (New York, 1990), 85, 103–4.

6. D.S.L. Cardwell, *Steam Power in the Eighteenth Century, A Case Study in the Application of Science* (London, 1963), quote 49.

7. Elijah Galloway, *History and Progress of the Steam Engine, with an extensive Appendix by Luke Herbert* (London, 1831), 51.

8. S. Schaffer, 'Making up Discovery', in M.A. Boden (ed.), *Dimensions of Creativity* (Cambridge, Massachusetts, 1994), 13–51.

9. Ian Inkster, 'Discoveries, Inventions and Industrial Revolutions: On the Varying Contributions of Technologies and Institutions from an International Historical Perspective', *History of Technology*, 18, (1996), 39–58, see especially 41–9.

10. Ian Inkster, *Science and Technology in History* (London, 1991), especially chs. 2, 3 and 4.

11. See appendix listings in Galloway *op. cit.* (7), 849–56.

Steam Power and Networks in China, 1860–98: The Historical Issues

NATHAN SIVIN AND Z. JOHN ZHANG

INTRODUCTION

Because the fate of steam power is often decided at the intersection of commerce and politics, finding 'useful and reliable knowledge' about it (to use Patrick O'Brien's phrase) calls for substantial attention to both.

Most historians seem to be certain that Chinese modernization before 1911 (as well as long afterward) was a failure, marked by rejection of key technologies, sporadic development of modern industry, meagre institutional innovation and reactionary government policies.

Attempts to account for these irrationalities have relied on several classic explanations.

One type concludes that traditional institutions, ideologies and habits of thought blocked the Chinese from responding positively to modern technology. It assumes that sensible people, in the absence of such impediments, would have leapt at the opportunity to change. Analyses of this kind sometimes appeal to the enormous gap between Confucian humanism and the imperatives of development, and sometimes to xenophobic attitudes that kept the governing class from recognizing where their advantage lay.

Another type argues that the imperialist powers' encroachments on Chinese sovereignty, in fact their goal of repackaging the empire into colonies, was bound to trap a weak government. Its always inferior technology doomed its efforts to reach the sustainable level of prosperity and military strength (*fuguoqiangbing* 富國強兵) that its statesmen dreamt of.

Both modes assume that the explanation does not vary with the technology – in other words, that when it comes to modernization we can treat technology as one big black box, and think of steam engines as all the same.

But there was no one response to steam power. The Chinese eagerly bought and began to build steamboats before 1875. In the same period, they firmly resisted great pressure to let nationals of the European powers construct railways. In making policies they treated steam locomotives as related not to ships but to telegraph lines.

Recently scholars have questioned whether Chinese moves on the complex international chess board of the late nineteenth century were in fact irrational. We take this as an important empirical question. We will ask why and how Chinese officials actually linked railway and telegraph technologies, how they weighed the options available to them, and whether the outcomes of their actions show that they were in fact misguided.

CIRCUMSTANCES

Before we proceed, we need to understand both the historical predicament of the Chinese state and the aims of the Western powers.[1]

The Manchus, a non-Han people, conquered China in the mid-seventeenth century and managed to rule it until 1911. But by 1865 the power of the ruling dynasty already had been fatally compromised, and much of its income had been bled away into foreign coffers.

When, from 1838 on, the imperial government seriously threatened the trade of British firms in opium, an important source of their profits, the UK government colluded with the traders to instigate a series of wars, and handily won them. The Chinese defeats had as much to do with institutions and social organization as with armaments. The result was indemnities so ruinous that, beginning in the 1850s, British, American and French functionaries collected China's customs duties directly; if anything was left over, they gave it to the government. The treaties that ended the Opium Wars also forced the Chinese to open their ports to foreign trade protected by foreign law.

From the late 1840s on, rebellions took a large portion of China's richest provinces out of imperial control for more than a decade. This turmoil further distracted the government and diminished its income from taxing its own people.

It is very much to the point that China as a whole did not experience so much as a year of peace and security from 1840 until 1949, or of even modest prosperity until the 1990s. In the second half of the nineteenth century, China could not exert sovereignty over parts of its own domains, and the foreign powers could back the government down on any point by threatening new hostilities.

Second, the European governments did not make such threats freely – a point that historians often overlook. Their imperative was not political subjugation but reliable commercial income, which is what colonies were for. Diplomacy and war were means to be used when trade was threatened, and in a way that did not threaten it even more. As the archives make clear, diplomats spent a great deal of effort calming down their own merchants in China who, eager to penetrate the interior, pushed unreasonably hard for expansion.

Third, to understand why the Chinese did what they did, it is helpful to avoid lumping them all together. Even if we consider only high officials, the fact is that they differed on almost every imaginable issue. On the one hand, many members of the imperial family, quite a few Manchus, and a number of highly placed Han Chinese were implacably opposed to accepting anything Western. They correctly understood an important point: occidental values, institutions and machines would corrode their own culture and their own authority. They thus invented the philosophy of 'just say no'.

On the other hand, European novelties had intrigued many others since the early seventeenth century, and some had learned that there were consequential ideas behind them. Extremely powerful officials such as Zeng Guofan 曾國藩 (1811–72), Li Hongzhang 李鴻章 (1823–1901), Zuo Zongtang 左宗棠 (1812–85) and Zhang Zhidong 張之洞 (1837–1909), learned the first great lesson of the Opium Wars, namely that the only way to defend oneself against foreigners was with their own technology. It was they who carried on the so-called Self-Strengthening Movement from about 1860 on, not only buying foreign machines but hiring Europeans to make them and to train their users. Li argued in 1874 that railway and telegraph lines were crucial to Chinese defence against naval attack, and got the Court's permission not long afterwards to build them for military purposes. Those purposes linked the two, for a reason that requires attention.

THE TECHNOLOGIES

Thomas P. Hughes showed in his *Networks of Power* that although railways, like steamship lines, use the power of steam for long-distance transportation, their patterns of evolution have nothing in common with those of steamships. On the other hand, they are very much like those of the telegraph, as well as those of the electrical grids that came along later.[2]

As a network technology, railways spread their steel ribbons in a continuous and highly visible way. They must have a public right of way, so a few troublemakers can easily paralyse them. They require a certain extent of standardization, and encourage more. As output grows its cost goes down, for instance as a train carries more passengers or freight. Those who invest in network technologies know that the first mile costs by far the most, that the first developer on the ground has an enormous advantage, that one needs some sort of local monopoly to make a profit, and that the way to increase one's gains is by extending the network.[3]

European entrepreneurs in China knew all that. In 1862 the Russian ambassador tried to establish a telegraph line between Peking and Tientsin. In 1863 foreign firms petitioned Li Hongzhang to let them build a railway between Shanghai and Suzhou.[4] It did not take Li and others who coped with these pressures long to understand the special character of the technologies involved, and to make decisions based on that under-standing. Let us look at an example.

THE WEDGE THEORY VS. THE SHANGHAI METHOD

The merchants knew that the crucial step was building a railway line a few hundred yards long, or terminating an undersea cable a few feet on shore. They thought of such an inconsequential beginning as a wedge that sooner or later would open wide the prospect of large networks. As good Victorians, they saw no possible rational objection to an enterprise that was bound to enrich them. What held the Chinese back, in their view, was ignorance of what was good for them. Entrepreneurs often reiterated their faith that a line, once in place, would sell itself.

The Chinese government was not at all ignorant. Officials who interacted with foreigners quickly became aware that the spread of networks, once begun, was inevitable. A directive from a central agency to the trade commissioner at Shanghai, in reply to a request by the consuls of nine nations to build an extremely modest telegraph line, states the point:

> It sounds quite proper. Yet, the evil intentions they harbour are not voiced. It should be clear that approval would pander to all the foreign ambassadors' desire to take an opportunity to get in, and one successful case will lead to demands for more. What they say about not taking this as a precedent ... is apparently bait.[5]

From the viewpoint of most Chinese officials, railways and telegraphs were means of invasion. They would threaten traditional livelihoods in regions already unstable, do violence to the relationship of man and nature, and in time of war transport enemy armies.

The European merchants' 'wedge theory' meant constant pressure from them to get permission for some innocuous-looking little specimen of engineering, and constant efforts to put one in place surreptitiously.[6] As the foreigners drove the wedge in bit by bit, the responsible Chinese officials encouraged their local subordinates not just to say no, as the reactionaries wanted, but to argue in a reasonable way that ordinary people would not stand for it. It did not, after all, take a horde of indignant labourers to cut a telegraph or rail line. To drive the point home, officials in the area of a line occasionally laid on just such a protest. This they came to call the 'Shanghai method', from where they first used it successfully to stop merchants' pressure.[7]

This was underhanded, like much that their opponents did. It was not xenophobia. The Chinese simply exploited the character of network technologies in an informed way to stop innovations that were out of their control, changes that they knew from experience would injure their own interests. At the same time, they were buying with alacrity arms and machines, even factories. These were no threat because, once the Chinese possessed them, their sellers could no longer control them. Self-strengtheners had no difficulty hiring foreign engineers for as long as they needed them to run the machinery and train workers.

The clash of interests came to a head in 1876. When a Shanghai–Suzhou railway proposal was turned down yet again, an American vice-consul organized a company to build what he announced would be a carriage road. When it was finished, it turned out to be a railway three-quarters of a mile long, with a locomotive small enough for 16 coolies to carry – not worth making a fuss over. A few months later the new line, already extended to cover four miles and with a larger engine, opened for passenger traffic. Since the government could neither destroy it nor stop its further spread without danger of another war, it quickly paid the astronomical price the company demanded, dismantled it, and moved it to the island of Taiwan.[8]

Why relocate it to an isolated backwater where it met no conceivable need? Historians have often described this decision as a typically irrational response to modernization. To the contrary, Taiwan was arguably the place where a network was least likely to propagate itself under foreign control in the 1870s. Despite an avalanche of indignation from foreign merchants, the Chinese officials knew what they were doing. The imperialist powers realized that further destabilizing a dysfunctional society could do long-term damage to commercial interests, so they would back only limited pressures. The merchants, having learned their lesson, stopped trying illegally to insert new wedges.

THE TELEGRAPH: NEW STRATAGEMS

Not only did this counter-intuitive move succeed, but the self-strengtheners quickly learned from the telegraph a positive way to manage foreign expansiveness.[9] In the early 1870s, various foreign companies pushed hard for contracts to build telegraph lines. Since the Chinese had no doctrine of territorial waters, they had not been interfering with the undersea cables that passed their shores on their way to link other continents. Rival Danish and British companies raced to connect their cables to land lines. The two companies, to make sure neither lost, signed a secret agreement not to compete and to split profits. The Danes illegally hauled a river cable onto land. When the Chinese government demanded in 1878 that they remove it, the counter-pressure from Europeans mounted. Li Hongzhang undercut this initiative by building his own telegraph line between Tientsin and Shanghai. By giving the politically powerless Danes (eager to be involved) a contract to operate it, he cut out the other powers.

The Chinese, with telegraph lines they owned and could control, had an unanswerable rationale for fending off foreign developers. They built new lines and extended existing ones with hired Danish help. Other European firms, competing feverishly for a piece of the action, could not unite against these projects. As Chinese officials built more land lines, they gained experience at playing off the rival interests of several countries against each other to keep them at loggerheads. By 1893 there were 167 telegraph sites across the empire. The Chinese had learned how to draw on modern expertise in the service of conventional statecraft.

CHINESE RAILWAY LINES

They duly applied this highly rational technology policy to railways.[10] Li Hongzhang built his first railway as part of a mining complex in 1882. After the Franco-Chinese war over Vietnam that year led to another humiliating loss, the Empress Dowager, who was no technocrat, gave Li permission to build another railway as part of a new defence plan, and in 1887 approved a long line.

Foreign capitalists cheered, because their help was obviously needed. They were avid for projects, since the main lines in Europe and the USA were finished, and there was a great depression of trade at home in the 1880s.

Historians have described the Chinese as improvident for putting the engineering in the hands of the British, whose high-quality, expensive, standardized practices they could barely afford. But from the viewpoint of the self-strengtheners, this forestalled bullying by all the other companies. The French or Germans simply could not, or would not, conform to the conservative British standards that the Chinese now insisted on. And the British, as secure sub-contractors, had less to bully China about.

But China's railway lines built from then until 1949 turned out to be a débâcle. The denouement came in 1895. When the war of that year with Japan ended, China was saddled with a foreign debt three times its annual revenues. It could no longer afford to reject demands by foreigners of one nationality after another to build and control new lines. Railways became just one more way for the foreign powers to mark off spheres of influence. The Chinese defended themselves with the methods they had learned, for instance by letting one power build in another's sphere of influence. But they could no longer say no. They were left with a many-gauged, almost useless patchwork that the greed of the Great Powers had assembled.

CONCLUSION

As China's loss of wealth, power and sovereignty accelerated from 1840 to 1911, it would be hard to show that a more considered set of decisions was possible. The Chinese archives show just as much careful, documented argument as do the papers of the British Foreign Office for the same period. That is not to assert that either side consistently made *wise* choices. In general, when the pressure was off, the Chinese statesmen in contact with foreigners tended to be open to new ideas. In times of crisis they tended to close down, as they still do – and as we still do.

A final question that we have to ponder after this whirlwind account is that of technological imperatives. It is a truism that network technologies, like other kinds, tend to satisfy needs that previously did not exist. They can do so only for peoples capable of changing their social practices to make room for new needs. As our example shows, the Chinese, despite the power of the reactionaries, did that to the extent that they could make the new systems serve their purposes rather than those of their enemies. But that extent was very limited in the second half of the nineteenth century, and not much greater in the first half of the twentieth. Steam engines do not dictate decisions. Action arises from human hopes, fears and interests.

To conclude, we would do well to look beyond abstractions such as steam power, and take as our unit not isolated inventions but the systems that use them. In order to understand why China's response to steam engines differed from that of, say, Bulgaria or Peru, we need to examine thoughtfully what the responses to particular kinds of steam engines, embodied in particular systems in their particular social circumstances, were.

It is a corollary of this argument that the pertinent elements of 'useful and reliable knowledge' were technological, political, economic and social, and that in this case – quite unlike others in China – knowledge of science happens to be beside the point.

CODA

Of the many nuggets that emerged in the discussions at this conference, one of the most striking was a point that Floris Cohen had made at a previous conference in this series. He was responding to an article in the so-called millennium edition of *The Economist*, which sought to account for the transformation in science and technology in the modern West by 'values, politics, and economic institutions' that were not combined in the same way elsewhere. That could not satisfy him as a historian, for it ignored the way in which systematic experiment and the quantification of phenomena came together to form something new and self-energizing. His current work attempts to describe what that something was. Whatever it was it was neither purely cognitive nor purely social.[11]

As an example we might consider the influence of Francis Bacon. When his name came up in our conversations, it was in reference to his empiricism, experimental, not at all quantitative, and oblivious to the uses of deduction. Attention to the empirical aspect alone overlooks the greatest source of his influence. In my reading, Bacon's *New Atlantis* argues that the authority of science can come to replace the waning authority of the Church, given the right organization and the right rituals. That would make it, for the first time, a truly collective enterprise. From its collective character would come secure social authority, greatly enhanced knowledge and enlarged dominion over Nature. His lawyerly narrow notion of method made his vision of the next steps in science far from percipient, but it is easy to see why his organizational insight inspired the Royal Society and the other burgeoning collectivities of the seventeenth century. It is only by looking at his convictions integrally that we can see what is unique about them.

I suggest, in other words, that what launched modern science was not two separate things, ideas and social relations, that may or may not interact in a given case, but one thing that historians divide into these categories to fit their professional specialisms. If we want to understand most adequately what that new something was, I suggest that we concentrate on framing our hypotheses in a language that is at the same time intellectual and social.[12]

On the other hand, I hope that we do not let a more sophisticated understanding of technical innovation distract us from the concrete, non-technical components of change. At several points in the discussion economic historians reminded us that it might be productive to think about the role of investment, and gave us a couple of excellent examples. But this topic never moved to the centre, because we tended to concentrate on more esoteric factors.

That is unfortunate. Desirable though it may be to clarify the constellations of Chinese and European abstractions and methodologies, we would do best to ask at the outset whether indeed they come into play in the fate of a given technology. With respect to some they are important; as for others, including network technologies, so far as I can see they played no significant role.

We are thus well advised to shift our attention elsewhere. The argument of this essay suggests that the keys to the response of the Chinese to network technologies were precisely the three to which *The Economist*'s article drew attention: values, politics and economics. I agree with Floris Cohen that alone they provide an inadequate basis for understanding historical change. But if we hope to reach some resolution on comparative questions, that does not mean we can leave them out of our toolbags as we concentrate on matters more familiar to orientalists and historians of science.

Notes and References

1. This essay draws on the reconstruction and interpretation of events in Zhang Zhong (now Prof. Z. John Zhang), 'The Transfer of Network Technologies to China, 1860–1898', Ph. D. dissertation, History and Sociology of Science, University of Pennsylvania, 1989. This study is based on sources in the archives of London, Paris, Beijing and Taipei. For one of the few well-balanced summaries of the historical background, see Ting-yee Kuo and Kwang-ching Liu, 'Self-strengthening: The Pursuit of Western Technology', in John K. Fairbank (ed.) *The Cambridge History of China*, vol. 10, part 1, (Cambridge, 1978), 491–542.

2. Thomas Parke Hughes, *Networks of Power: Electrification in Western Society, 1880–1930* (Baltimore, 1983).

3. Zhang, *op cit.* (1), xii–xiii.

4. Zhang, *op cit.* (1), 3–4.

5. Zhang, *op cit.* (1), 14, translated from Jindaishi Yenjiusuo (Institute of Modern History, Academia Sinica), *Hai fang dang* (Maritime defense archives), section on telegraphs (Taipei, 1957), 77 (slightly modified).

6. On their conscious use of the word 'wedge' see the *North-China Herald*, February 16 1867.

7. *Hai fang dang* (see note 5 above), 17, 19–20. For the first application of the term, in 1865, see Zhang, *op cit.* (1), 11.

8. Zhang, *op cit.* (1), 18–26.

9. For details on the topic of this section, see Zhang, *op cit.* (1), ch. 3, 84–129.

10. For particulars see Zhang, *op cit.* (1), chs. 4–5, 130–257.

11. H. Floris, Cohen, 'Inside Newcomen's Fire Engine, or: The Scientific Revolution and the Rise of the Modern World' (unpublished paper for conference on 'Regimes for the Generation of Useful and Reliable Knowledge in Europe and Asia 1368–1815', Cumberland Lodge, Windsor Great Park, 14–16 April 2000); Anonymous, 'The Road to Riches', *The Economist*, 31 December 1999, 10–12.

12. For an example see Geoffrey Lloyd and Nathan Sivin, *The Way and the Word. Science and Medicine in Early China and Greece* (New Haven, 2002).

A Comment on Sivin and Zhang

R. BIN WONG

Sivin and Zhang's essay usefully informs us about the political context within which certain steam-powered technologies were introduced into nineteenth-century China. In addition, as historians of science they reliably recount the particular features of network technologies. Finally, because they are also historians of China conversant with the relevant archival records, they are able to show us how Chinese decisions regarding telegraph and railroad development were reasonable and understandable, if only partially successful at best. The surprise of their essay lies in their conclusion that their examples affirm the *Economist* article about the importance of 'values, politics and economics' to explain the development and application of science and technology in the past two centuries. By my reading of the *Economist* article, the 'values', 'politics' and 'economic institutions' this piece invokes do not correspond closely to the factors at work in the story that Sivin and Zhang tell us.

The *Economist* article 'The Road to Riches' argues that the increases in incomes and rising standards of living over the past two and half centuries is a remarkable event in the history of the human species. The author tells us how it came to be that 'Western Europe and its American offshoot' forged these mighty changes. He rejects the idea that European science and technology on their own can supply an explanation of the dramatic economic changes. Going beyond a focus on science and technology the *Economist* essay explains that Europeans had the proper set of 'values' which included 'acquisitiveness' and 'selflessness', themes made famous by Max Weber and associated by him with Protestantism. Places where people lack these values, not surprisingly, did not create the road to riches. The argument proves a perennial as it has sprouted up in a similar form in the writing of numerous policy-oriented social scientists, well presented in the volume edited by Lawrence Harrison and Samuel Huntington, entitled *Culture matters: how values shape human progress*. The correct 'politics' mentioned in the *Economist* article are those it associates with pluralism and rights to property harking back to the Magna Carta signed in 1215 between King John and the English barons. The 'economic institutions' it summons include the familiar ones of contract law, patents and company laws. In brief, the 'values, politics, and economic institutions' of 'the road to riches' that are used to explain economic growth and the application of science and technology to achieve this end were created in and by the West.

However effective an explanation these 'values, politics and economic institutions' provide for Western economic growth, and that large subject is certainly open to discussion and debate, it by no means follows logically that in order to take advantage of the growth-building techniques first formulated in the West, other places need adopt the same complement of 'values, politics and economic institutions'. Furthermore, it is not empirically true that the absence of such material progress can be best explained by the absence of the factors claimed significant for the initial formulation of economic growth in the West. Indeed, Sivin and Zhang help us understand that it wasn't because the Chinese lacked 'acquisitiveness' or 'selflessness' that they had problems developing network technologies. Indeed, they were no less rational and sensible than Europeans, though they were less successful in achieving their intended results. One of the basic constraints late nineteenth-century Chinese officials faced, a constraint far more binding in the case of railroads than telegraphs, was imposed by the interests and power of Western businessmen and their governments. This is a 'politics' grounded in differences of power that Sivin and Zhang use to explain the limited success of railroad development in China. Chinese officials may have failed to develop adequate economic institutions to guide the development of railroads but this stemmed in part at least from their political inability to control the demands of late nineteenth-century foreigners. Certainly political weakness was a far more immediate constraint that prevented effective negotiations with foreign interests than was absence of 'acquisitiveness', 'pluralism' or European-style contract law.

A truly amazing set of economic changes took place in the nineteenth century. The initial changes were unmistakably rooted in European and at times specifically British conditions. Explaining those developments remains a serious challenge that is in key ways distinct from explaining why the technologies crucial to growth developed in Europe but did not spread more evenly and easily around the world. Regarding this second problem, the answers that Sivin and Zhang give us for the somewhat different experiences of the telegraph and railroad in late nineteenth-century China make clear that the difficulties that the Chinese encountered were more directly produced by a clash of interests between Chinese leaders and Western businessmen and decided by their relative power more than by any set of cultural differences distinguishing Chinese from Westerners.

Contents of Former Volumes

* indicates out of print

IAN INKSTER, Pursuing Big Books: Technological Change in Global History.
HANS-JOACHIM BRAUN, History of Technology in Germany: A Success Story?

ANNA GUAGNINI, Patent Agents, Legal Advisers and Guglielmo Marconi's Breakthrough in Wireless Telegraphy.
JOSÉ M. ORTIZ-VILLAJOS, International Patenting in Spain Before the Civil War.